Synthesis Lectures on Ocean Systems Engineering

Series Editor

Nikolas Xiros, University of New Orleans, New Orleans, LA, USA

The series publishes short books on state-of-the-art research and applications in related and interdependent areas of design, construction, maintenance and operation of marine vessels and structures as well as ocean and oceanic engineering.

Fidaa Karkori · Alexander Arnfinn Olsen

Inventory of Hazardous Materials

Fidaa Karkori ⓘ
Southampton, UK

Alexander Arnfinn Olsen
Southampton, UK

ISSN 2692-4420   ISSN 2692-4471 (electronic)
Synthesis Lectures on Ocean Systems Engineering
ISBN 978-3-031-76733-3   ISBN 978-3-031-76734-0 (eBook)
https://doi.org/10.1007/978-3-031-76734-0

This Springer imprint is published by the registered company Springer Nature Switzerland AG
The registered company address is: Gewerbestrasse 11, 6330 Cham, Switzerland

If disposing of this product, please recycle the paper.

Preface

In May 2009, the Hong Kong International Convention for the Safe and Environmentally Sound Recycling of Ships (Ship Recycling Convention) was formally adopted at a Diplomatic Conference in Hong Kong. The purpose of the Ship Recycling Convention is to prevent, reduce, minimise and, to the extent practicable, eliminate accidents, injuries and other adverse effects on human health and the environment caused by ship recycling, and to enhance a vessel's safety, protection of human health and the environment throughout a vessel's operating life (Article 1). Specific regulations for safe and environmentally sound recycling of vessels are annexed to the Convention. In order to adopt procedures for facilitating the effective implementation of the Convention, IMO has been focusing its efforts in developing a set of guidelines for the development of the inventory of hazardous materials, development of a ship recycling plan, authorisation of ship recycling facilities, safe and environmentally sound ship recycling, etc.

Essential to the implementation scheme of the Convention is the development and maintenance of a document referred to as the Inventory of Hazardous Materials ("Inventory" or "IHM"), which was previously known as the "Green Passport" (GP) in IMO Resolution A.962(23), IMO Guidelines on Ship recycling. The Inventory is vessel-specific and covers the whole life of the vessel, from construction, throughout the vessel's operating life up to the time of preparation for scrapping at the end of the vessel's useful life. The main differences between an Inventory and a Green Passport are that compared to a Green Passport, an Inventory requires a more detailed and reasonably accurate account of the listed hazardous substances in the inventory and the additional procedure of sampling where required, to be carried out for existing vessels.

Since the introduction of the Class Guide for the Class Notation Green Passport (GP) in March 2011, several revisions to the guidelines have been adopted by IMO, these revisions have further assisted in a more accurate preparation of the Inventory. This Class Guide for the Inventory of Hazardous Materials supersedes the Class Guide for the Class Notation Green Passport (GP), 2011 (Updated February 2014).

This text provides the Class requirements for the review and verification of the initial Inventory for new construction vessels and existing vessels and also for the maintenance and verification of the Inventory for a vessel in service. The initial inventory covers Part I of the IHM, as it is the only part that needs to be maintained during the life of the vessel. The preparation and review of Parts II and III of the IHM are outside the scope of this text. As the requirements in this text are aligned to the Guidelines developed by IMO for the Hong Kong Convention, the IHM notation will enable issuance of "The International Certificate on Inventory of Hazardous Materials" when the Hong Kong Convention enters into force. The parts in this text which are additional classification requirements and Class interpretations of the aforementioned Ship Recycling Convention and/or Resolution are presented in non-italics "Times New Roman" type style. Under these parts are also, wherever applicable, Class interpretations of IMO, IACS, and other related instruments.

The application of this text is optional. However, recognising the increasing attention of the maritime community on the protection of the marine environment and occupational health and safety relating to ship recycling, which is further supported by Regulation (EU) No 1257/2013 of the European Parliament and of the Council on ship recycling (EU SRR), the compliance with which is considered as a route for early ratification of the Ship Recycling Convention, designers, shipbuilders, shipowners, and operators are encouraged to apply this text.

Southampton, UK   Fidaa Karkori
                                                                   Alexander Arnfinn Olsen

Acknowledgements The authors would like to express their grateful thanks to Dr. Dieter Merkle at Springer, Boopalan Renu at Straive, and Drs. Anders Widt and Phillipe Sauvin for their knowledge and support in the development of this text.

This title is published with the kind permission of the American Bureau of Shipping.

Contents

List of Figures

List of Tables

General

1.1 General

The importance of an inventory detailing the type, amount, and location of hazardous materials used in a vessel's construction and operations is increasingly recognised as a means to enhance onboard safety and environmental awareness, both throughout the ship's economic life and at the end of the vessel's useful life, when the ship is being prepared for recycling. This text has been developed with the objective of assisting designers, equipment suppliers, shipbuilders, ship repair facilities, operators and owners in the practical and reasonable formulation and maintenance of this inventory.

1.2 Applications

This textbook is applicable to new and existing vessels for which an Inventory has been submitted for review and verification to the satisfaction of Class, thereby enabling issuance of the notation IHM, or Class-specific equivalent. Obtaining this notation will assist owners and operators to comply with Regulations 4, 5 and 10 of the Hong Kong International Convention for the Safe and Environmentally Sound Recycling of Ships, 2009 (SR/CONF/45), as amended.

1.3 Scope

This textbook has been developed with the objective of promoting the industry's ongoing commitment to providing safe working conditions of the vessel's crew, protecting the marine environment and in recognition of the maritime community's efforts, as a stake

© The Author(s), under exclusive license to Springer Nature Switzerland AG 2025
F. Karkori and A. A. Olsen, *Inventory of Hazardous Materials*, Synthesis Lectures on Ocean Systems Engineering, https://doi.org/10.1007/978-3-031-76734-0_1

holder, assisting in the safe and environmentally sound recycling of vessels. The text includes information on inventorying hazardous materials on new or existing vessels, conducting verification surveys and the issuance and maintenance of the Class notation IHM, or Class-specific equivalent. This text covers the Class requirements for the review and verification survey of Part I, the hazardous and potentially hazardous materials in a vessel's structure and equipment as detailed in the Inventory. Parts II and III, covering the operationally generated waste and stores are not required to be completed and hence are not subject to review until a vessel is in the process of being prepared for recycling.

1.4   Notations

At the request of the owner, vessels which have had their Inventory reviewed and verified in accordance with the guidelines outlined in this text, and which are confirmed to the satisfaction of Class Survey, may receive the Class notation IHM, or Class-specific equivalent. The requirements outlined in this text for the purpose of obtaining the Class notation IHM, or Class-specific equivalent, may be considered as additional to all other relevant requirements of the Class Rules. Compliance with the applicable requirements of the following IMO documents, specifically with regard to the inventory of hazardous materials, is typically a prerequisite:

(1) IMO SR/CONF/45, Hong Kong International Convention for the Safe and Environmentally Sound Recycling of Ships, 2009, in particular with the following Regulations: Regulation 4—Controls of ship's hazardous materials; Regulation 5—Inventory of Hazardous Materials: paragraphs 1–3, Regulation 5—Inventory of Hazardous Materials: paragraphs 1–3
   • Regulation 4—Controls of ship's hazardous materials
   • Regulation 5—Inventory of Hazardous Materials: paragraphs 1–3
   • Regulation 10—Surveys: subparagraphs 1.1–1.3; and
(2) IMO Resolution MEPC.269(68), 2015 Guidelines for the Development of the Inventory of Hazardous Materials.

1.5   Issuing of the IHM (or Class-Specific Equivalent) Notation

The IHM notation, or Class-specific equivalent, is available to vessels contracted for construction on or after the effective date of this text. The IHM notation, or Class-specific equivalent, is also available to existing vessels subject to compliance with the guidance provided at Chap. 3, Sect. 1.4, para. 4.2, along with the Class requirements specified in Chap. 3, Sect. 1.4, para. 4.3.

1.6 Maintenance of GP Notation and Conversion from GP to IHM Notation (or Class-Specific Equivalents)

Those vessels that currently possess a GP notation may continue to remain eligible to maintain the GP notation, provided the requirements contained within the Class Rules for the Class Notation Green Passport (GP) continues to be satisfied. Vessels under construction may receive a new Class notation, IHM, provided the GP notation was requested prior to the date the revised IHM Notation became effective. A vessel is "under construction," for purposes of this provision, if the contract date for construction between the shipbuilder and the prospective owner is prior to the date the IHM Notation became effective. Existing vessels that have a GP Notation may receive an IHM notation provided:

- These vessels have Inventories that have been prepared using the procedure for "new vessels" in compliance with Chap. 3, Sect. 4.1 of this text, or
- These existing vessels which currently maintain an Inventory prepared by using the procedure for "existing vessels" in compliance with Chap. 3, Sect. 4.2 of this text, along with their Inventory's being reviewed and revised where necessary by a Class approved service supplier. Refer to Chap. 3, Sect. 4.2 of this text.

Note: All references to the Green Passport (GP) Notation and Inventory of Hazardous Materials (IHM) Notation are for benefit of brevity; always refer to the Class-specific equivalents for currency.

1.7 Future Regional or Governmental Regulations

Individual Flag Administrations may in the future have in place additional requirements pertaining to the requirements for mandatorily maintaining an Inventory. When such additional requirements are to be enforced this text will be revised accordingly, considering such additional requirements. Further reference may be made to Sect. 1.8.

1.8 EU Ship Recycling Regulation

On 30 December 2013, the European Union adopted Regulation (EU) No 1257/2013 of the European Parliament and of the Council on ship recycling, amending Regulation (EC) No 1013/2006 and Directive 2009/16/EC. Unlike the Hong Kong Convention, which allowed new installations containing hydrochlorofluorocarbons (HCFCs) until 1 January 2020, the EU Flagged ships Regulation (EU) No 1257/2013 does not permit such a relaxation. To enable ease of understanding the implementation dates of the European Union Regulation No 1257/2013, reference is to be made to Table 1.1.

Table 1.1 Comparison TimeLine between Regulation EU No 1257/2013 and the HK convention

	Date of entry in to force of Regulation EU No 1257/2013: 30 December 2013		30 December 2013 SR/CONF/45, Hong Kong International Convention
Applicability to ships	Specific stage of ship	Timeline for the development of Part I of an IHM	Timeline for the development of Part I of an IHM
EU Flagged New Ships	"**New ship**" means a ship: (1) for which the building contract is placed on or after the application date of this Regulation; or (2) in the absence of a building contract, the keel of which is laid, or which is at a similar stage of construction on or after six months after the application date of this Regulation; or (3) the delivery of which is on or after 30 months after the application date of this Regulation **Art. 3(2)**	Date of application: The application of the Regulation will be from the earlier of the following two dates (1) 6 months after the date that the combined maximum annual ship recycling output of the ship recycling facilities included in the European List constitutes not less than 2.5 million light displacement tonnes (LDT) (2) on 31 December 2018 **Art. 32(1)**	• The HK Convention will enter into force only 24 months after the date of ratification by at least 15 states • The 15 states should represent a combined merchant fleet of at least 40% of the gross tonnage of the world's merchant shipping and whose combined maximum annual ship recycling volume during the preceding 10 years constitutes not less than three per cent of the gross tonnage of the combined merchant shipping of the same states

(continued)

Table 1.1 (continued)

	Date of entry in to force of Regulation EU No 1257/2013: 30 December 2013		30 December 2013 SR/CONF/45, Hong Kong International Convention
Applicability to ships	Specific stage of ship	Timeline for the development of Part 1 of an IHM	Timeline for the development of Part 1 of an IHM
EU Flagged Existing Ships	"**Existing ship**" means a ship which is not a new ship	From 31 December 2020 For ships going for recycling an IHM is to be prepared prior recycling Once the European List of Recycling Facilities is issued, EU Flagged Vessels may only be recycled at facilities listed in the EU List [The European List is to be published not later than 31 December 2016] **Art. 32(2);Art. 5(2) SP-2; Art 16(2)**	
Non-EU Flagged Ship	Is a ship flying the flag of a third country (i.e. non-EU) when calling at a port or anchorage of a Member State **Art. 12(1)**	From 31 December **Art. 12(1)**	

Documents to Be Submitted to Class

2

2.1 Plans and Documentation

In this chapter, Tables 2.1 and 2.2 show the documents to be submitted to the Class Engineering Office for review or information and those required to be placed onboard for the initial survey of the Inventory for new and existing vessels, respectively. Table 2.3 shows the documents to be made available onboard for Annual Surveys and Table 2.4 shows the documents to be submitted to the Class Engineering Office and placed onboard when an additional survey is requested for the update of Part I of the IHM.

2.2 Engineering Review

Plans submitted to an Engineering office will be reviewed for compliance based on the following:

2.2.1 Development of Part I of the Inventory for New Vessels

(1) The Inventory was prepared at the design and construction stage in accordance with Chap. 3, Sect. 4.1 of this text
(2) The Inventory corresponds to the information indicated on the Material Declaration and Supplier's Declaration of Conformity
(3) The Inventory identifies hazardous materials contained in ship structure, equipment and coating, along with the location and approximate quantity
(4) Hazardous materials listed in Appendix 1, Table A1.1 of this text are not used onboard the vessel above their threshold values, unless permitted by the Convention

© The Author(s), under exclusive license to Springer Nature Switzerland AG 2025
F. Karkori and A. A. Olsen, *Inventory of Hazardous Materials*, Synthesis Lectures on Ocean Systems Engineering, https://doi.org/10.1007/978-3-031-76734-0_2

Table 2.1 Submissions for the development of Part I of the inventory for new vessels

Type of document	Description	For review (R) or information (I)	Placed onboard ship for survey (S)

Inventory of Hazardous Materials (IHM)	• Hazardous materials listed in Appendices 1 and 2 meeting Regulation 4 and 5.1 of SR/CONF/45 and Chap. 3, Sect. 4.1 of this text
	• In standard format as per Appendix 2 of this text
	• Cover page of the IHM to include the following details of the ship:
	  – Ship name
	  – IMO number
	  – International call sign
	  – Flag state
	  – Port of registry
	  – Date of register with the flag state
	  – Ship type
	  – Shipyard and Hull No
	  – Class ID
	  – Ship owner name and address
		R	S

Material Declaration (MD)	One Material Declaration for one product using IMO standard Material Declaration form in Appendix 6 of this text. Material Declaration should be prepared for all products regardless of the existence of hazmat specified in Appendix 1, Tables A1.1 and A1.2 of this text	I	

(continued)

Table 2.1 (continued)

Type of document	Description	For review (R) or information (I)	Placed onboard ship for survey (S)
Supplier's Declaration of Conformity (SDoC)	Every Material Declaration is to be accompanied by a Supplier's Declaration of Conformity. A single Supplier's Declaration of Conformity may refer to several items of conformity declarations as long as each item and its related product information and supplementary information can be clearly identified. Refer to Appendix 7 of this text	I	
Location Diagram of Hazmat	Diagrams to show the location of materials listed in Appendix 1, Table A1.1. Refer to Appendices 1 and 5 of this text	R	S
Ship Specific Plans	General arrangement plan or details of different locations of the ship	I	
Quality management system—procedures to safeguard the proper updating of the IHM	• Details of designated person responsible for maintaining and updating the IHM • Records of new installations, repairs, maintenance, modifications to ship • Archive of all the associated documentation • Random sampling policy, as applicable	I	S

Note *The collected documents should be listed in an archive which should follow the ship throughout its operational life, and it may be in electronic format*

Table 2.2 Submissions for the Development of Part I of the inventory for existing vessels

Type of document	Description	For review (R) or information (I)	Placed onboard ship for survey (S)

Inventory of Hazardous Materials (IHM)	• Hazardous Materials listed in Appendices 1 and 2 meeting Regulation 4 and 5.2 of SR/CONF/45 and Chap. 3, Sect. 4.2 of this text
	• In standard format as per Appendix 2 of this text
	• Cover page of the IHM to include the following details of the ship:
	  – Ship name
	  – IMO number
	  – International call sign
	  – Flag state
	  – Port of registry
	  – Date of register with the flag state
	  – Ship type
	  – Shipyard and Hull No
	  – Class ID
	  – Ship owner name and address
		R	S

Material Declaration (MD)	• Where possible • One Material Declaration for one product using IMO standard Material Declaration form in Appendix 6 of this text. Material Declaration should be prepared for all products regardless of the existence of hazmat specified in Appendix 1, Tables A1.1 and A1.2 of this text	I	

(continued)

Table 2.2 (continued)

Type of document	Description	For review (R) or information (I)	Placed onboard ship for survey (S)
Supplier's Declaration of Conformity (SDoC)	• Where possible • Every Material Declaration is to be accompanied by a Supplier's Declaration of Conformity. A single Supplier's Declaration of Conformity may refer to several items of conformity declaration as long as each item and its related product information and supplementary information can be clearly identified. Refer to Appendix 7 of this text	I	
Visual/Sampling Check Plan	Refer to the examples provided in Appendix 5 of this text	I	S

(continued)

Table 2.2 (continued)

Type of document	Description	For review (R) or information (I)	Placed onboard ship for survey (S)

Report of Visual/Sampling Check	Is to comprise the following:
	• Visual/Sampling check plan (refer to the example of a check plan in Appendix 5 of this text
	• Checklist covering:
	  – At least, items in the indicative list in Appendix 5 of this text
	  – Considered spaces, areas, structures, equipment, fittings, as specific to the ship; and
	  – Updated with the results of onboard check and laboratory sample analysis results
	• Support documents used for the development of visual/sampling check report (e.g., ship files, vendor product specifications, Material Declarations, laboratory analysis report, etc.)
		I	

Location Diagram of Hazmat	Diagrams to show the location of materials listed in Appendix 1, Table A1.1, refer to Appendices 1 and 5 of this text	I	

(continued)

Table 2.2 (continued)

Type of document	Description	For review (R) or information (I)	Placed onboard ship for survey (S)
Ship Specific Plans	General arrangement plan or details of different locations of the ship	I	
Quality management system—procedures to safeguard the proper updating of the IHM	• Details of designated person responsible for maintaining and updating the IHM • Records of new installations, repairs, maintenance, modifications to ship • Archive of all the associated documentation • Random sampling policy, as applicable	I	S

Note The collected documents should be listed in an archive which should follow the ship throughout its operational life, and it may be in electronic format

Table 2.3 Annual survey of Part I of the inventory

Type of document	Description	Placed onboard ship for survey (S)
Inventory of Hazardous Materials (IHM)	IHM with IHM supplements if applicable	S
Material Declaration (MD)	Material Declaration collected for purchases of materials, machinery or equipment, coating renewal and spares from the date of the last approval of the IHM or IHM supplements thereof	S
Supplier's Declaration of Conformity (SDoC)	Supplier's Declaration of Conformity supporting Material Declaration collected from the date of the last approval of the IHM or IHM supplements thereof	S
Location Diagram of Hazmat	Updated diagrams, where applicable	S
Quality management system—procedures to safeguard the proper updating of the IHM	• Records of new installations, repairs, maintenance, modifications to ship • Records of changes to inventory with fields for dates and signatures	S

Note The collected documents should be listed in an archive which should follow the ship throughout its operational life, and it may be in electronic format

(5) Hazardous materials listed in Appendix 1, Table A1.2 of this text, when used onboard the ship above their threshold values, are reported in the Inventory

(6) The Inventory contains diagrams showing the locations of materials listed in Appendix 1, Table A1.1 of this text

(7) A quality management system is established to safeguard the quality and continuity of the IHM which includes the identity of the designated person, a system for maintaining and updating of the Inventory, records of new installations, repairs, maintenance and modifications to a ship or ship's IHM designated person and records of changes to the Inventory. Proper maintenance of an archive of all the associated documentation should also be included and it should include that new installations of equipment, repairs, and refitting are accompanied by a Material Declaration and the Supplier's Declaration of Conformity, as provided by the suppliers of parts and equipment delivered. There may also be a random sampling policy for new or existing ship; and

Table 2.4 Submissions when additional survey of Part I of the inventory is requested

Type of document	Description	For review (R) or information (I)	Placed onboard ship for survey (S)
Inventory of Hazardous Materials (IHM)	IHM with IHM supplements	R	S
Material Declaration (MD)	Material Declaration collected for purchases of materials, machinery or equipment, coating renewal and spares from the date of the last approval of the IHM or IHM supplements thereof	R	
Supplier's Declaration of Conformity (SDoC)	Supplier's Declaration of Conformity supporting Material Declaration collected from the date of the last approval of the IHM or IHM supplements thereof	R	
Location Diagram of Hazmat	Updated diagrams, where applicable	R	S
Quality management system—procedures to safeguard the proper updating of the IHM	• Records of new installations, repairs, maintenance, modifications to ship • Records of changes to inventory with fields for dates and signatures	I	S

Note The collected documents should be listed in an archive which should follow the ship throughout its operational life, and it may be in electronic format

(8) A software tool may be used to support the IHM development and maintenance process and the management of all the relevant documents, information and data.

2.2.2 Development of Part I of the Inventory for Existing Vessels

(1) The Inventory was prepared in accordance with Chap. 3, Sect. 4.2 of this text:
  (a) By a company in the Class list of approved service suppliers; or

(b) By the shipowner with the assistance of a company from the Class list of approved service supplier (the company engaged has signed off on the visual/sampling check plan and the report of visual/sampling check)

(2) The Inventory corresponds with the visual/sampling check report

(3) The Inventory identifies hazardous materials complying with this chapter, Sect. 2.1(3)

(4) The classification of items as "potentially containing hazardous material (PCHM)" are noted in the remarks column of the Inventory where applicable.

(5) Hazardous materials listed in Appendix 1, Table A1.1 of this text, where present in quantities above the respective threshold values, are reported in the Inventory.

(6) Hazardous materials listed in Appendix 1, Table A1.2 of this text, complying with this chapter, Sect. 2.1(4)

(7) The Inventory includes diagrams complying with this chapter, Sect. 2.1(5)

(8) A quality management system is established to safeguard the quality and continuity of the IHM which includes the identity of the designated person, a system for maintaining and updating of the Inventory, records of new installations, repairs, maintenance and modifications to a ship or ship's IHM designated person and records of changes to the Inventory; and

(9) A software tool may be used to support the IHM development and maintenance process and the management of all the relevant documents, information and data.

2.2.3 When an Additional Survey of Part I of the Inventory is Requested

(1) The Inventory has been updated in accordance with Chap. 3, Sect. 4.3 of this text

(2) The Inventory corresponds with the records of new installations, repairs, maintenance, modifications to ship and Material Declarations

(3) Hazardous materials listed in Appendix 1, Table A1.1 of this text, complying with this chapter, Sect. 2.1(4)

(4) Hazardous materials listed in Appendix 1, Table A1.2 of this text, complying with this chapter, Sect. 2.1(5); and

(5) The Deletion of equipment and/or parts of the ship's structure previously classed as PCHM in the Inventory were supported with reasons and evidence that the equipment, system and/or area do not contain hazardous materials or which have subsequently been sampled and found to contain hazardous material, which have been identified in the revised Inventory or a supplement to the Inventory thereof.

2.3 Surveys

For the requirements for surveys, please refer to Chap. 5 for guidance.

Development of the Inventory of Hazardous Materials

3

Note: The following text is an extract from IMO Resolution MEPC.269(68), 2015—Guidelines for the Development of the Inventory of Hazardous Materials.

3.1 Introduction

3.1.1 Objectives of the Guidelines

These guidelines provide recommendations for developing the Inventory of Hazardous Materials (hereinafter referred to as "the Inventory" or "the IHM") to assist compliance with regulation 5 (Inventory of Hazardous Materials) of the Hong Kong International Convention for the Safe and Environmentally Sound Recycling of Ships, 2009 (hereinafter referred to as "the Convention").

3.1.2 Application

These guidelines have been developed to provide relevant stakeholders (e.g. shipbuilders, equipment suppliers, repairers, shipowners and ship management companies) with the essential requirements for the practical and logical development of the Inventory.

© The Author(s), under exclusive license to Springer Nature Switzerland AG 2025
F. Karkori and A. A. Olsen, *Inventory of Hazardous Materials*, Synthesis Lectures
on Ocean Systems Engineering, https://doi.org/10.1007/978-3-031-76734-0_3

3.1.3   Objectives of the Inventory

The objectives of the Inventory are to provide ship-specific information on the actual hazardous materials present on board, in order to protect health and safety and to prevent environmental pollution at ship recycling facilities. This information will be used by the ship recycling facilities in order to decide how to manage the types and amounts of materials identified in the Inventory of Hazardous Materials (regulation 9 of the Convention).

3.2   Definitions

The terms used in these guidelines have the same meaning as those defined in the Convention, with the following additional definitions which apply to these guidelines only.

2.1 **Exemption** as referred to in regulation 5 of the Convention) means materials specified in paragraph 3.3 in these guidelines that do not need to be listed on the IHM, even if such materials or items exceed the IHM threshold values.

2.2 **Fixed** means the conditions that equipment or materials are securely fitted with the ship, such as by welding or with bolts, riveted or cemented, and used at their position, including electrical cables and gaskets.

2.3 **Homogeneous material** means a material of uniform composition throughout that cannot be mechanically disjointed into different materials, meaning that the materials cannot, in principle, be separated by mechanical actions such as unscrewing, cutting, crushing, grinding and abrasive processes.

2.4 **Loosely fitted equipment** means equipment or materials present on board the ship by the conditions other than "fixed," such as fire extinguishers, distress flares, and lifebuoys.

2.5 **Product** means machinery, equipment, materials and applied coatings on board a ship.

2.6 **Supplier** means a company which provides products, which may be a manufacturer, trader or agency.

2.7 **Supply chain** means the series of entities involved in the supply and purchase of materials and goods, from raw materials to final product.

2.8 **Threshold value** is defined as the concentration value in homogeneous materials.

Hong Kong International Convention for the Safe and Environmentally Sound Recycling of Ships, 2009

Article 2
Definitions

1. "**Convention**" means the Hong Kong International Convention for the Safe and Environmentally Sound Recycling of Ships, 2009.
2. "**Administration**" means the Government of the State whose flag the ship is entitled to fly, or under whose authority it is operating.
3. "**Competent Authority(ies)**" means a governmental authority, or authorities designated by a Party as responsible, within specified geographical area(s) or area(s) of expertise, for duties related to Ship Recycling Facilities operating within the jurisdiction of that Party as specified in this Convention.
4. "**Organisation**" means the International Maritime Organisation.
5. "**Secretary-General**" means the Secretary-General of the Organisation.
6. "**Committee**" means the Marine Environment Protection Committee of the Organisation.
7. "**Ship**" means a vessel of any type whatsoever operating or having operated in the marine environment and includes submersibles, floating craft, floating platforms, self-elevating platforms, Floating Storage Units (FSUs), and Floating Production Storage and Offloading Units (FPSOs), including a vessel stripped of equipment or being towed.
8. "**Gross tonnage**" means the gross tonnage (GT) calculated in accordance with the tonnage measurement regulations contained in Annex I to the International Convention on Tonnage Measurement of Ships, 1969, or any successor convention.
9. "**Hazardous Material**" means any material or substance which is liable to create hazards to human health and/or the environment.
10. "**Ship Recycling**" means the activity of complete or partial dismantling of a ship at a Ship Recycling Facility in order to recover components and materials for reprocessing and re-use, whilst taking care of hazardous and other materials, and includes associated operations such as storage and treatment of components and materials on site, but not their further processing or disposal in separate facilities.
11. "**Ship Recycling Facility**" means a defined area that is a site, yard or facility used for the recycling of ships.
12. "**Recycling Company**" means the owner of the Ship Recycling Facility or any other organisation or person who has assumed the responsibility for operation of the Ship Recycling activity from the owner of the Ship Recycling Facility and who on assuming such responsibility has agreed to take over all duties and responsibilities imposed by this Convention.

3.3   Requirements for the Inventory

3.3.1   Scope of the Inventory

The Inventory consists of:
   Part I: Materials contained in ship structure or equipment
   Part II: Operationally generated wastes; and
   Part III: Stores.

3.3.2   Materials to Be Listed in the Inventory

3.2.1 Appendix 1 of these guidelines (Items to be listed in the Inventory of Hazardous Materials), provides information on the hazardous materials that may be found on board a ship. Materials set out in Appendix 1 should be listed in the Inventory. Each item in Appendix 1 of these guidelines is classified under Tables A1-1, A1-2, A1-3 or A1-4, according to its properties:

   (1) Table A1-1 comprises the materials listed in Appendix 1 of the Convention
   (2) Table A1-2 comprises the materials listed in Appendix 2 of the Convention
   (3) Table A1-3 (Potentially hazardous items) comprises items which are potentially hazardous to the environment and human health at ship recycling facilities; and
   (4) Table A1-4 (Regular consumable goods potentially containing hazardous materials) comprises goods which are not integral to a ship and are unlikely to be dismantled or treated at a ship recycling facility.

3.2.2 Tables A1-1 and A1-2 correspond to Part I of the Inventory. Table A1-3 corresponds to Parts II and III and Table A1-4 corresponds to Part III.

3.2.3 For loosely fitted equipment, there is no need to list this in Part I of the Inventory. Such equipment which remains on board when the ship is recycled should be listed in Part III.

3.2.4 Those batteries containing lead acid or other hazardous materials that are fixed in place should be listed in Part I of the Inventory. Batteries that are loosely fitted, which includes consumer batteries and batteries in stores, should be listed in Part III of the Inventory.

3.2.5 Similar materials or items that contain hazardous materials that potentially exceed the threshold value can be listed together (not individually) on the IHM with their general location and approximate amount specified there (hereinafter referred to as "bulk listing"). An example of how to list those materials and items is shown in row 3 of Table A3-1 of Appendix 3.

3.3.3 Exemptions—Materials not Required to Be Listed in the Inventory

3.3.1 Materials listed in Table A1-2 that are inherent in solid metals or metal alloys, such as steels, aluminium, brasses, bronzes, plating and solders, provided they are used in general construction, such as hull, superstructure, pipes or housings for equipment and machinery, are not required to be listed in the Inventory.

3.3.2 Although electrical and electronic equipment is required to be listed in the Inventory, the amount of hazardous materials potentially contained in printed wiring boards (printed circuit boards) installed in the equipment does not need to be reported in the Inventory.

3.3.4 Standard Format of the Inventory of Hazardous Materials

The Inventory should be developed on the basis of the standard format set out in Appendix 2 of these guidelines: Standard format of the Inventory of Hazardous Materials. Examples of how to complete the Inventory are provided for guidance purposes only.

3.3.5 Revision to Threshold Values

Revised threshold values in Tables A1-1 and A1-2 of Appendix 1 should be used for IHMs developed or updated after the adoption of the revised values and need not be applied to existing IHMs and IHMs under development. However, when materials are added to the IHM, such as during maintenance, the revised threshold values should be applied and recorded in the IHM.

3.4 Requirements for Development of the Inventory

3.4.1 Development of Part I of the Inventory for New Ships[1]

4.1.1 **Part I of the Inventory for new ships should be developed at the design and construction stage.**

4.1.2 **Checking of materials listed in Table A1-1**

During the development of the Inventory (Part I), the presence of materials listed in Table A1-1 of Appendix 1 should be checked and confirmed; the quantity and location of Table A1-1 materials should be listed in Part I of the Inventory. If such materials are used in compliance with the Convention, they should be listed in Part I of the Inventory.

Any spare parts containing materials listed in Table A1-1 are required to be listed in Part III of the Inventory.

4.1.3 Checking of materials listed in Table A1-2

If materials listed in Table A1-2 of Appendix 1 are present in products above the threshold values provided in Table A1-2, the quantity and location of the products and the contents of the materials present in them should be listed in Part I of the Inventory. Any spare parts containing materials listed in Table A1-2 are required to be listed in Part III of the Inventory.

4.1.4 Process for checking of materials

The checking of materials as provided in paragraphs 4.1.2 and 4.1.3 above should be based on the Material Declaration furnished by the suppliers in the shipbuilding supply chain (e.g. equipment suppliers, parts suppliers, material suppliers).

To preclude the introduction of non-compliant components into a vessel's structure or equipment after the inventory for the vessel has been prepared, Material Declarations are also to be obtained for purchases of spare parts that would be included in the vessel's spares at delivery. If any of the Material Declarations for spare parts contain materials listed in Appendix 1, Table A1-1 or A1-2 of this text above the respective threshold values, these spare parts are to be documented in an appendix to Part 1 of the IHM. When these spare parts are used, Part 1 of the IHM is to be updated accordingly.

Note
[1] in ascertaining whether a ship is a "new ship" or an "existing ship" according to the Convention, the term "a similar stage of construction" in regulation 1.4.2 of the annex to the Convention means the stage at which:

(a) construction identifiable with a specific ship begins; and
(b) assembly of that ship has commenced comprising at least 50 tonnes or 1% of the estimated mass of all structural material, whichever is less.

3.4.2   Development of Part I of the Inventory for Existing Ships

4.2.1 In order to achieve comparable results for existing ships with respect to Part I of the Inventory, the following procedure should be followed:

(1) collection of necessary information

(2) assessment of collected information

(3) preparation of visual/sampling check plan

(4) onboard visual check and sampling check; and

(5) preparation of Part I of the Inventory and related documentation.

4.2.2 The determination of hazardous materials present on board existing ships should, as far as practicable, be conducted as prescribed for new ships, including the procedures described in sections 6 and 7 of these guidelines. Alternatively, the procedures described in this section may be applied for existing ships, but these procedures should not be used for any new installation resulting from the conversion or repair of existing ships after the initial preparation of the Inventory.

4.2.3 The procedures described in this section should be carried out by the shipowner, who may draw upon expert assistance. Such an expert or expert party should not be the same as the person or organisation authorised by the Administration to approve the Inventory.

To assist shipowners, identify experts for the development of Part I of the IHM, the majority of classification societies have developed a qualification scheme to qualify companies offering this expertise. Where such a scheme exists, Class will usually qualify companies based on the following documented procedures, unless stipulated otherwise:

- Procedures for the development of Part I of the IHM, in compliance with Hong Kong Convention and Resolution MEPC 269(68)—2015 Guidelines for the development of IHM; and
- Qualification and training of expert

For existing vessels to qualify for an Class IHM notation, or Class-specific equivalent, Class may require shipowners to engage a company from the classification society's own preferred list of approved service suppliers for the development of Part 1 of the IHM or if shipowners wish to develop the inventory in-house, they may be required to engage a company from the classification society's own preferred list of service suppliers to carry out visual/sampling checks onboard the vessel (an expert from the company engaged is to sign off on the visual/sampling check plan and the report of visual/sampling check).

4.2.4 Reference is made to Appendix 4 (Flow diagram for developing Part I of the Inventory for existing ships) and Appendix 5 (Example of development process for part I of the Inventory for existing ships).

4.2.5 Collection of necessary information (step 1)

The shipowner should identify, research, request and procure all available documentation regarding the ship. Information that will be useful includes maintenance, conversion and repair documents; certificates, manuals, ship's plans, drawings and technical specifications; product information data sheets (such as Material Declarations); and hazardous material inventories or recycling information from sister ships. Potential sources of information could include previous shipowners, the ship builder, historical societies, classification society records and ship recycling facilities with experience working with similar ships.

4.2.6 Assessment of collected information (step 2)

The information collected in step 1 above should be assessed. The assessment should cover all materials listed in Table A1-1 of Appendix 1; materials listed in Table A1-2 should be assessed as far as practicable. The results of the assessment should be reflected in the visual/sampling check plan.

4.2.7 Preparation of visual/sampling check plan (step 3)

4.2.7.1 To specify the materials listed in Appendix 1 of the guidelines, a visual/sampling check plan should be prepared considering the collated information and any appropriate expertise. The visual/sampling check plan should be based on the following three lists:
(1) List of equipment, system and/or area for visual check (any equipment, system and/or area specified regarding the presence of the materials listed in Appendix 1 by document analysis should be entered in the List of equipment, system and/or area for visual check)
(2) List of equipment, system and/or area for sampling check (any equipment, system and/or area which cannot be specified regarding the presence of the materials listed in Appendix 1 by document or visual analysis should be entered in the List of equipment, system and/or area as requiring sampling check. A sampling check is the taking of samples to identify the presence or absence of hazardous material contained in the equipment, systems, and/or areas, by suitable and accepted methods such as laboratory analysis); and

(3) List of equipment, system and/or area classed as "potentially containing hazardous material" (any equipment, system and/or area which cannot be specified regarding the presence of the materials listed in Appendix 1 by document analysis may be entered in the List of equipment, system and/or area classed as "potentially containing hazardous material" without the sampling check. The prerequisite for this classification is a comprehensible justification such as the impossibility of conducting sampling without compromising the safety of the ship and its operational efficiency).

4.2.7.2 Visual/sampling checkpoints should be all points where:
(1) the presence of materials to be considered for the Inventory Part I as listed in Appendix 1 is likely
(2) the documentation is not specific; or
(3) materials of uncertain composition were used.

To preclude the introduction of non-compliant components into a vessel's structure or equipment after the inventory for the ship is prepared, the visual/sampling check plan is to include checking of spare parts onboard the ship that may be used for the vessel's structure or equipment. If any of the spare parts for the vessel's structure or equipment onboard the ship are found to contain materials listed in Appendix 1, Table A1-1 or A1-2 of this text above their respective threshold values, these spar e parts are to be documented in an appendix to Part I of the IHM. When these spare parts are used, Part I of the IHM is to be updated accordingly.

4.2.8 Onboard visual/sampling check (step 4)

4.2.8.1 The onboard visual/sampling check should be carried out in accordance with the visual/sampling check plan. When a sampling check is carried out, samples should be taken, and the sample points should be clearly marked on the ship plan and the sample results should be referenced. Materials of the same kind may be sampled in a representative manner. Such materials are to be checked to ensure that they are of the same kind. The sampling check should be carried out drawing upon expert assistance. Photographic evidence is to be included along with the check and assignment of materials that are considered to be of the same kind for the purpose of taking representative samples.

4.2.8.2 Any uncertainty regarding the presence of hazardous materials should be clarified by a visual/sampling check. Checkpoints should be documented in the ship's plan and may be supported by photographs.

4.2.8.3 If the equipment, system and/or area of the ship are not accessible for a visual check or sampling check, they should be classified as "potentially containing hazardous material."

The prerequisite for such classification should be the same prerequisite as outlined in Chap. 4, Sect. 2.7.

Any equipment, system and/or area classed as "potentially containing Hazardous Material" may be investigated or subjected to a sampling check at the request of the shipowner during a later survey (e.g. during repair, refit or conversion).

4.2.9 Preparation of Part I of the Inventory and related documentation (step 5)

If any equipment, system and/or area is classed as either "containing hazardous material" or "potentially containing hazardous material," their approximate quantity and location should be listed in Part I of the Inventory. These two categories should be indicated separately in the "Remarks" column of the Inventory.

4.2.10 Testing methods

4.2.10.1 Samples may be tested by a variety of methods. "Indicative" or "field tests" may be used when:
(1) the likelihood of a hazard is high
(2) the test is expected to indicate that the hazard exists; and
(3) the sample is being tested by "specific testing" to show that the hazard is present.

4.2.10.2 Indicative or field tests are quick, inexpensive and useful on board the ship or on site, but they cannot be accurately reproduced or repeated, and cannot identify the hazard specifically, and therefore cannot be relied upon except as "indicators."

4.2.10.3 In all other cases, and in order to avoid dispute, "specific testing" should be used. Specific tests are repeatable, dependable and can demonstrate definitively whether a hazard exists or not. They will also provide a known type of the hazard. The methods indicated are found qualitative and quantitative appropriate and only testing methods to the same effect can be used. Specific tests are to be carried out by a suitably accredited laboratory, working to international standards[2] or equivalent, which will provide a written report that can be relied upon by all parties.

Note
[2] For example, ISO 17025.

4.2.10.4 Specific test methods for Appendix 1 materials are provided in Appendix 9.

4.2.11 Diagram of the location of hazardous materials on board a ship. Preparation of a diagram showing the location of the materials listed in Table A is recommended in order to help ship recycling facilities gain a visual understanding of the Inventory. The location diagram is a necessary submission that is to be included with other documents submitted for review.

3.4.3   Maintaining and Updating Part I of the Inventory During Operations

4.3.1 Part I of the Inventory should be appropriately maintained and updated, especially after any repair or conversion or sale of a ship. The maintenance of Part I of the Inventory is to be based on Material Declarations furnished by suppliers for equipment, parts, material etc. Material Declarations are to be obtained for all purchases that may impact a ship's structure and equipment (e.g. materials, machinery or equipment, spare parts, etc.).

If any of the Material Declarations contain materials listed in Tables A1-1 or A1-2 of Appendix 1of this text above the respective threshold values, Part I of the Inventory is to be updated accordingly with the use of these materials, equipment or spare parts onboard the ship

4.3.2 Updating of Part I of the Inventory in the event of new installation

If any machinery or equipment is added to, removed or replaced or the hull coating is renewed, Part I of the Inventory should be updated according to the requirements for new ships as stipulated in paragraphs 4.1.2 to 4.1.4. Updating is not required if identical parts or coatings are installed or applied.

4.3.3 Continuity of Part I of the Inventory

Part I of the Inventory should belong to the ship and the continuity and conformity of the information it contains should be confirmed, especially if the flag, owner or operator of the ship changes.

3.4.4 Development of Part II of the Inventory (Operationally Generated Waste)

4.4.1 Once the decision to recycle a ship has been taken, Part II of the Inventory should be developed before the final survey, taking into account that a ship destined to be recycled shall conduct operations in the period prior to entering the Ship Recycling Facility in a manner that minimises the amount of cargo residues, fuel oil and wastes remaining on board (regulation 8.2 of the Convention).

4.4.2 **Operationally Generated Wastes to Be Listed in the Inventory**

If the wastes listed in Part II of the Inventory provided in Table A1-3 (Potentially hazardous items) of Appendix 1 are intended for delivery with the ship to a ship recycling facility, the quantity of the operationally generated wastes should be estimated, and their approximate quantities and locations should be listed in Part II of the Inventory.

3.4.5 Development of Part III of the Inventory (Stores)

4.5.1 Once the decision to recycle has been taken, Part III of the Inventory should be developed before the final survey, considering the fact that a ship destined to be recycled shall minimise the wastes remaining on board (regulation 8.2 of the Convention). Each item listed in Part III should correspond to the ship's operations during its last voyage.

4.5.2 Stores to Be Listed in the Inventory

If the stores to be listed in Part III of the Inventory provided in Table A1-3 of Appendix 1 are to be delivered with the ship to a ship recycling facility, the unit (e.g. capacity of cans and cylinders), quantity and location of the stores should be listed in Part III of the Inventory.

4.5.3 Liquids and gases sealed in ship's machinery and equipment to be listed in the Inventor

If any liquids and gases listed in Table A1-3 of appendix 1 are integral in machinery and equipment on board a ship, their approximate quantity and location should be listed in Part III of the Inventory. However, small amounts of lubricating oil, anti-seize compounds and grease which are applied to or injected into machinery and equipment to maintain normal performance do not fall within the scope of this provision. For subsequent completion of Part III of the Inventory during the recycling preparation processes, the quantity of liquids and gases listed in Table A1-3 of Appendix 1 required for normal operation, including the related pipe system volumes, should be prepared and documented at the design and construction stage.

This information belongs to the ship, and continuity of this information should be maintained if the flag, owner or operator of the ship changes.

Where information is available pertaining to the liquids and gases listed in Appendix 1, Table A1-3, which are integral within the machinery and equipment on board the ship, their approximate quantities and locations are to be listed in an appendix to Part I of the IHM to facilitate the preparation of Part III of the IHM during the preparation of the ship for recycling.

4.5.4 Regular consumable goods to be listed in the inventory

Regular consumable goods, as provided in Table A1-4 of Appendix 1 should not be listed in Part I or Part II but should be listed in Part III of the Inventory if they are to be delivered with the ship to a Ship Recycling Facility. A general description including the name of item (e.g. TV set), manufacturer, quantity and location should be entered in Part III of the Inventory. The check on materials provided for in paragraphs 4.1.2 and 4.1.3 of these guidelines does not apply to regular consumable goods.

3.4.6   Description of Location of Hazardous Materials on Board

The locations of hazardous materials on board should be described and identified using the name of location (e.g. second floor of engine-room, bridge DK, APT, No.1 cargo tank, frame number) given in the plans (e.g. general arrangement, fire and safety plan, machinery arrangement or tank arrangement).

3.4.7   Description of Approximate Quantity of Hazardous Materials

In order to identify the approximate quantity of hazardous materials, the standard unit used for hazardous materials should be kg, unless other units (e.g. m^3 for materials of liquid or gases, m^2 for materials used in floors or walls) are considered more appropriate. An approximate quantity should be rounded up to at least two significant figures.

3.5   Requirements for Ascertaining the Conformity of the Inventory

3.5.1   Design and Construction Stage

The conformity of Part I of the Inventory at the design and construction stage should be ascertained by reference to the collected Supplier's Declaration of Conformity described in Sect. 7 and the related Material Declarations collected from suppliers.

3.5.2   Operational Stage

Shipowners should implement the following measures in order to ensure the conformity of Part I of the Inventory:

(1) to designate a person as responsible for maintaining and updating the Inventory (the designated person may be employed ashore or on board)
(2) the designated person, in order to implement paragraph 4.3.2, should establish and supervise a system to ensure the necessary updating of the Inventory in the event of new installation
(3) to maintain the Inventory including dates of changes or new deleted entries and the signature of the designated person
(4) to provide related documents as required for the survey or sale of the ship; and
(5) to verify that new installations of equipment, repairs and refitting are accompanied by a Material Declaration and the Supplier's Declaration of Conformity, as provided by the suppliers of parts and equipment delivered.

3.6   Material Declaration

3.6.1   General

Suppliers to the shipbuilding industry should identify and declare whether or not the materials listed in Table A1-1 or Table A1-2 are present above the threshold value specified in Appendix 1 of these guidelines. However, this provision does not apply to chemicals which do not constitute a part of the finished product.

3.6.2   Information Required in the Declaration

6.2.1 At a minimum the following information is required in the Material Declaration:
    (1) date of declaration
    (2) Material Declaration identification number
    (3) supplier's name
    (4) product name (common product name or name used by manufacturer)
    (5) product number (for identification by manufacturer)
    (6) declaration of whether or not the materials listed in Table A1-1 and Table A1-2 of Appendix 1 of these guidelines are present in the product above the threshold value stipulated in Appendix 1 of these guidelines; and
    (7) mass of each constituent material listed in Table A1-1 and/or Table A1-2 of Appendix 1 of these guidelines if present above threshold value.

6.2.2 An example of the Material Declaration is shown in Appendix 6.

The submission of an MSDS for hazardous material is not acceptable in lieu of a Material Declaration. MSDS may be accepted as a supplement to a Material Declaration.

3.7   Supplier's Declaration of Conformity

3.7.1   Purpose and Scope

7.1.1 The purpose of the Supplier's Declaration of Conformity is to provide assurance that the related Material Declaration conforms to Sect. 6.2, and to identify the responsible entity.

7.1.2 The Supplier's Declaration of Conformity remains valid as long as the products are present on board.

7.1.3 The supplier compiling the Supplier's Declaration of Conformity should establish a company policy.[3]

The company policy on the management of the chemical substances in products which the supplier manufactures or sells should cover:

Note
[3] A recognised quality management system may be utilised.

(1) Compliance with law:

The regulations and requirements governing the management of chemical substances in products should be clearly described in documents which should be kept and maintained; and

(2) Obtaining of information on chemical substance content:

In procuring raw materials for components and products, suppliers should be selected following an evaluation, and the information on the chemical substances they supply should be obtained.

3.7.2   Contents and Format

7.2.1 The Supplier's Declaration of Conformity should contain the following:
  (1) unique identification number
  (2) name and contact address of the issuer
  (3) identification of the subject of the Declaration of Conformity (e.g. name, type, model number, and/or other relevant supplementary information)
  (4) statement of conformity
  (5) date and place of issue; and
  (7) signature (or equivalent sign of validation), name and function of the authorised person(s) acting on behalf of the issuer.

7.2.2 An example of the Supplier's Declaration of Conformity is shown in Appendix 7.

3.8   List of Appendices

Appendix 1   Items to be listed in the Inventory of Hazardous Materials
Appendix 2   Standard format of the Inventory of Hazardous Materials
Appendix 3   Example of the development process for part I of the Inventory for new ships
Appendix 4   Flow diagram for developing Part I of the Inventory for existing ships
Appendix 5   Example of the development process for Part I of the Inventory for existing ships
Appendix 6   Form of Material Declaration
Appendix 7   Form of Supplier's Declaration of Conformity
Appendix 8   Examples of Tables A1-1 and A1-2 materials of Appendix 1 with CAS-numbers
Appendix 9   Specific test methods
Appendix 10   Examples of radioactive sources

Maintenance and Continuity of Part I of the Inventory

4

4.1   Introduction

To maintain the IHM notation, or Class-specific equivalent, it is the responsibility of the shipowner to keep the Inventory updated at all times. Whenever there are new installations (machinery, equipment, hull coating) added to the vessel that includes materials listed in Appendix 1 of this text, as permitted by the Hong Kong Convention, the Inventory is to be updated where applicable. The shipowner is to request Class for an additional survey if there are significant amount of changes to the Inventory. Refer to Chap. 2, Table 2.1D for documents to be submitted to the Class Engineering Office for review and Chap. 5, Sect. 7 for requirements of an additional survey. New installations containing hazardous materials listed in Appendix 1, Table A1.1 are prohibited, or restricted except for installations containing hydrochlorofluorocarbons (HCFCs), which were permitted up to 1 January 2020. The alternative procedure, following the five steps in Chap. 3, Sect. 4.2.1 of this text to develop the Inventory for existing vessels is not to be used for updating the Inventory in the event of new installations. Refer to Chap. 3, Sect. 4.2.2 of this text.

4.2   Maintenance of the IHM

A maintenance manual for the Inventory complying with Chap. 2, Sect. 4.2/3 (7), with relevant dates and the signature of the designated person is to be kept up to date. Refer to Chap. 3, Sect. 4.3 and Chap. 3, Sect. 5.2. The deletion of equipment, system and/or area previously classed as "potentially containing hazardous materials (PCHM)" is to comply with Chap. 2, Sect. 3.5(6). The shipowner is to maintain that the Inventory is maintained with current information.

© The Author(s), under exclusive license to Springer Nature Switzerland AG 2025
F. Karkori and A. A. Olsen, *Inventory of Hazardous Materials*, Synthesis Lectures
on Ocean Systems Engineering, https://doi.org/10.1007/978-3-031-76734-0_4

4.3   Continuities of the IHM

Where there is a change of flag, owner or operator of the vessel, the contents of the Inventory and supporting documents and maintenance manual are to be confirmed as containing the latest information and passed on to the next shipowner. Refer to Chap. 3, Sect. 4.3.3. A new designated person responsible for the Inventory and system to maintain and update the Inventory is to be established by the subsequent owner or operator of the vessel.

Survey of Part I of the Inventory

5.1   Introduction

The Inventory is to be subjected to surveys for the issuance and maintenance of the IHM notation.

5.2   Initial Survey

Refer to Chap. 2, Tables 2.1A and B for the list of documents to be made available onboard for the initial survey of the Inventory for new and existing ships, respectively. The initial survey is to be carried out with an Inventory that has been reviewed by Class Engineering without outstanding technical comments. The survey is to verify that the Inventory, especially the location of hazardous materials, is consistent with the arrangements, structure and equipment of the vessel through an onboard visual inspection. Upon completion of the initial survey, the Class optional notation IHM would be granted.

5.3   Annual Survey

The Inventory will be subjected to an annual survey in accordance with the Class Rules for survey after construction in the course of completing other annual and periodical surveys. Refer to Chap. 2, Table 2.1C for the list of documents to be made available onboard for the annual survey of the Inventory. The annual survey is to verify the following:

© The Author(s), under exclusive license to Springer Nature Switzerland AG 2025
F. Karkori and A. A. Olsen, *Inventory of Hazardous Materials*, Synthesis Lectures
on Ocean Systems Engineering, https://doi.org/10.1007/978-3-031-76734-0_5

- The Inventory has been maintained and updated to reflect changes in vessel structure and equipment based on the records of new installations, repairs, maintenance, modifications to ship, to the satisfaction of the Surveyor
- Material Declaration and Supplier's Declaration of Conformity have been collected for purchases of materials, machinery or equipment, coating renewal and spares from the date of the last Survey verification of the Inventory or Inventory supplements thereof. Deletion of equipment and/or parts of the ship's structure previously classed as PCHM from the Inventory complies with the requirements of Chap. 2, Sect. 3.3.5; and
- The Inventory, especially the location of hazardous materials, is consistent with the arrangements, structure and equipment of the vessel through an onboard visual inspection.

5.4   Additional Surveys

When a ship undergoes a replacement or repair of the structure, equipment, systems, fittings, arrangements or material, which has a significant impact on the ship's Inventory, the shipowner is to request Class for an additional survey of the Inventory. Refer to Chap. 2, Table 2.1D for the list of documents to be made available onboard for the additional survey of Inventory. The additional survey is to be carried out with an updated Inventory or an Inventory supplement that has been reviewed by Engineering without outstanding technical comments. The survey is to verify that the updated Inventory or additional supplements to the Inventory, especially the location of hazardous materials, is consistent with the arrangements, structure and equipment of the vessel through an onboard visual inspection.

Appendix A: Items to Be Listed in the Inventory of Hazardous Materials

Appendix A.1, A.2, A.3 and A.4.

F. Karkori and A. A. Olsen, *Inventory of Hazardous Materials*, Synthesis Lectures
on Ocean Systems Engineering, https://doi.org/10.1007/978-3-031-76734-0

Table A.1 Materials listed in Appendix 1 of the Annex to the Convention

No	Materials		Inventory			Threshold value
			Part I	Part II	Part III	
A-1	Asbestos		·			0.1%[4]
A-2	Polychlorinated Biphenyls (PCBs)		·			50 mg/kg[5]
A-3	Ozone Depleting Substances*	CFCs	·			No threshold Value[6]
		Halons	·			
		Other fully halogenated CFCs	·			
		Carbon Tetrachloride	·			
		1,1,1-Trichloroethane (Methyl Chloroform)	·			
		Hydrochlorofluorocarbons (HCFCs) (b)	·			
		Hydrobromofluorocarbons	·			
		Methyl Bromide	·			
		Bromochloromethane	·			
A-4	Anti-fouling systems containing organotin compounds as a biocide		·			2,500 mg total tin/kg[7]

Note The Hong Kong Convention allowed new installations containing hydrochlorofluorocarbons (HCFCs) until 1 January 2020. The product contacting HCFC should be listed in the Inventory.

[4] In accordance with regulation 4 of the Convention, for all ships, new installation of materials which contain asbestos shall be prohibited. According to the UN recommendation "Globally Harmonised System of Classification and Labelling of Chemicals (GHS)" adopted by the United Nations Economic and Social Council's Sub-Committee of Experts on the Globally Harmonised System of Classification and Labelling of Chemicals (UNSCEGHS), the UN's Sub-Committee of Experts, in 2002 (published in 2003), carcinogenic mixtures classified as Category 1A (including asbestos mixtures) under the GHS are required to be labelled as carcinogenic if the ratio is more than 0.1%. However, if 1% is applied, this threshold value should be recorded in the Inventory and, if available, the Material Declaration and can be applied not later than five years after the entry into force of the Convention. The threshold value of 0.1% need not be retroactively applied to those Inventories and Material Declarations.

[5] In accordance with regulation 4 of the Convention, for all ships, new installation of materials which contain PCBs shall be prohibited. The Organisation set 50 mg/kg as the threshold value referring to the concentration level at which wastes, substances and articles containing, consisting of or contaminated with PCB are characterised as hazardous under the Basel Convention.

[6] "No threshold value" is in accordance with the Montreal Protocol for reporting ODS. Unintentional trace contaminants should not be listed in the Material Declarations and in the Inventory.

[7] This threshold value is based on the Guidelines for brief sampling of anti-fouling systems on ships (resolution MEPC.104[49])

Table A.2 Materials listed in Appendix 2 of the Annex to the Convention

No	Materials	Inventory			Threshold Value
		Part I	Part II	Part III	
B-1	Cadmium and Cadmium Compounds	·			100 mg/kg[8]
B-2	Hexavalent Chromium and Hexavalent Chromium Compounds	·			1,000 mg/kg[8]
B-3	Lead and Lead Compounds	·			1,000 mg/kg[8]
B-4	Mercury and Mercury Compounds	·			1,000 mg/kg[8]
B-5	Polybrominated Biphenyl (PBBs)	·			50 mg/kg[9]
B-6	Polybrominated Diphenyl Ethers (PBDEs)	·			1,000 mg/kg[8]
B-7	Polychloronaphthalanes (more than 3 chlorine atoms)	·			50 mg/kg[10]
B-8	Radioactive Substances	·			No threshold value[11]
B-9	Certain Short chain Chlorinated Paraffins (Alkanes, C10-C13, chloro)	·			1%[12]

Note

[8] The Organisation set this as the threshold value referring to the Restriction of Hazardous Substances (RoHS Directive 2011/65/EU, Annex II)

[9] The Organisation set 50 mg/kg as the threshold value referring to the concentration level at which wastes, substances and articles containing, consisting of or contaminated with PBB are characterised as hazardous under the Basel Convention

[10] The Organisation set 50 mg/kg as the threshold value referring to the concentration level at which wastes, substances and articles containing, consisting of or contaminated with PCN are characterised as hazardous under the Basel Convention

[11] All radioactive sources should be included in the Material Declaration and in the Inventory. Radioactive source means radioactive material permanently sealed in a capsule or closely bonded and in a solid form that is used as a source of radiation. This includes consumer products and industrial gauges with radioactive materials. Examples are listed in appendix 10

[12] The Organisation set 1% as the threshold value referring to the EU legislation that restricts Chlorinated Paraffins from being placed on the market for use as substances or as constituents of other substances or preparations in concentrations higher than 1% (EU Regulation 1907/2006, Annex XVII Entry 42 and Regulation 519/2012)

Table A.3 Potentially hazardous items

No	Properties		Goods	Inventory		
				Part I	Part II	Part III
C-1	Liquid	Oiliness	Kerosene			·
C-2			White Spirit			·
C-3			Lubricating Oil			·
C-4			Hydraulic Oil			·
C-5			Anti-seize Compounds			·
C-6			Fuel Additives			·
C-7			Engine Coolant Additives			·
C-8			Antifreeze Fluids			·
C-9			Boiler and Feed Water Treatment and Test Re-agents			·
C-10			De-ioniser Regenerating Chemicals			·
C-11			Evaporator Dosing and Descaling Acids			·
C-12			Paint Stabilisers/Rust Stabilisers			·
C-13			Solvents/Thinners			·
C-14			Paints			·
C-15			Chemical Refrigerants			·
C-16			Battery Electrolyte			·
C-17			Alcohol, Methylated Spirits			·
C-18	Gas	Explosives and Inflammables	Acetylene			·
C-19			Propane ·			·
C-20			Butane			·
C-21			Oxygen			·
C-22		Greenhouse Gases	CO2			·
C-23			Perfluorocarbons (PFCs)			·
C-24			Methane			·
C-25			Hydrofluorocarbon (HFCs)			·
C-26			Nitrous Oxide (N2O)			·
C-27			Sulphur Hexafluoride (SF6)			·
C-28	Liquid	Oiliness	Bunkers: Fuel Oil			·
C-29			Grease			·
C-30			Waste Oil (Sludge)		·	

(continued)

Table A.3 (continued)

No	Properties		Goods	Inventory		
				Part I	Part II	Part III
C-31			Bilge and/or Waste-Water Generated by the After-treatment Systems Fitted on Machinery		.	
C-32			Oily Liquid Cargo Tank Residues		.	
C-33			Ballast Water		.	
C-34			Raw Sewage		.	
C-35			Treated Sewage		.	
C-36			Non-oily Liquid Cargo Residues		.	
C-37	Gas	Explosibility and Inflammability	Fuel Gas		.	
C-39	Solid		Dry Cargo Residues		.	
C-40			Medical Waste/Infectious Waste		.	
C-41			Incinerator Ash13		.	
C-42			Rubbish		.	
C-43			Fuel Tank Residues		.	
C-44			Oily Solid Cargo Tank Residues		.	
C-45			Oily or Chemical Contaminated Rags		.	
C-46			Batteries (incl. lead acid batteries)			.
C-47			Pesticides/Insecticide Sprays			.
C-48			Extinguishers			•
C-49			Chemical Cleaner (incl. Electrical Equipment Cleaner, Carbon Remover)			•
C-50			Detergent/Bleach (could be a liquid)			•
C-51			Miscellaneous Medicines			•
C-52			Firefighting Clothing and Personal Protective Equipment			•
C-53			Dry Tank Residues		•	
C-54			Cargo Residues		•	
C-55			Spare parts which contain materials listed in Table A or Table B			•

Note
[13] Definition of rubbish is identical to that in MARPOL Annex V. However, incinerator ash is classified separately because it may include hazardous substances or heavy metals

Table A.4 Regular consumable goods potentially containing hazardous materials[14]

No	Properties	Examples	Inventory Part I	Part II	Part III
D-1	Electrical and electronic equipment	Computers, refrigerators, printers, scanners, television sets, radio sets, video cameras, video recorders, telephones, consumer batteries, fluorescent lamps, filament bulbs, lamps			●
D-2	Lighting equipment	Fluorescent lamps, filament bulbs, lamps			●
D-3	Non-ship specific furniture, interior and similar equipment	Chairs, sofas, tables, beds, curtains, carpets, garbage bins, bed-linen, pillows, towels, mattresses, storage racks, decoration, bathroom installations, toys, not structurally relevant or integrated artwork			●

Note
[14] This table does not include ship-specific equipment integral to ship operations, which has to be listed in Part I of the inventory

Appendix B: Standard Format of the Inventory of Hazardous Materials[15]

This Appendix contains the standard format of the Inventory of Hazardous Materials (IHM).

Note

[15] Examples of how to complete the Inventory are provided for guidance purposes only in accordance with paragraph 3.4 of the guidelines.

Part 1 Hazardous materials contained in the ship's structure and equipment

I-1 Paints and coating systems containing materials listed in table a and Table B of Appendix 1

No	Application of paint	Name of paint	Location	Materials (classification in Appendix 1)	Approx. quantity	Remarks
1	Anti-drumming compound	Primer, xx Co., xx primer #300	Hull part	Lead	35.00 kg	
2	Anti-fouling	xx Co., xx coat #100	Underwater parts	TBT	120.00 kg	

© The Editor(s) (if applicable) and The Author(s), under exclusive license to Springer Nature Switzerland AG 2025
F. Karkori and A. A. Olsen, *Inventory of Hazardous Materials*, Synthesis Lectures on Ocean Systems Engineering, https://doi.org/10.1007/978-3-031-76734-0

I-2 Equipment and machinery containing materials listed in table a and Table B of Appendix 1

No	Application of equipment and machinery	Location	Materials (classification in Appendix 1)	Parts where used	Approx. quantity	Remarks
1	Switch board	Engine control room	Cadmium	Housing coating	0.02 kg	
			Mercury	Heat gauge	<0.01 kg	less than 0.01 kg
2	Diesel engine	Engine room	Lead	Starter for blower	0.01 kg	Revised by XXX in Oct 2008 (revoking No.2)
3	Diesel generator (× 3)	Engine room	Lead	Ingredient of copper compounds	0.01 kg	
4	Radioactive level gauge	No. 1 cargo tank	Radioactive substances	Gauge	5 (1.8 E + 11) $Ci_{(Bq)}$	Radionuclides: ^{60}Co

I-3 Structure and hull containing materials listed in table a and Table B of Appendix 1

No	Application of structural element	Location	Materials (classification in Appendix 1)	Parts where used	Approx. quantity	Remarks
1	Wall panel	Accommodation	Asbestos	Insulation	2,500.00 kg	
2	Wall insulation	Engine control room	Lead	Perforated Plate	0.01 kg	
			Asbestos	Insulation	25.00 kg	

Part II

Operationally generated waste No	Location[1]	Name of item (classification in Appendix 1) and detail (if any) of the item	Approximate quantity	Remarks
1	Rubbish locker	Rubbish (food waste)	35.00 kg	
2	Bilge tank	Bilgewater	15.00 m^3	

(continued)

(continued)

Operationally generated waste No	Location[1]	Name of item (classification in Appendix 1) and detail (if any) of the item	Approximate quantity	Remarks
3	No.1 cargo hold	Dry cargo residues (iron ore)	110.00 kg	
4	No.2 cargo hold	Waste oil (sludge) (crude)	120.00 kg	
5	No.1 ballast tank	Ballast water	2,500.00 m^3	
		Sediments	250.00 kg	

Note

[1] The location of a part II or part III item should be entered in order based on its location, from a lower level to an upper level and from a fore part to an aft part. The location of part I items is recommended to be described similarly, as far as practicable.

3 Part III

Stores
III-1 Stores

No	Location[1]	Name of item (classification in Appendix 1)	Unit quantity		Figure		Approx. quantity m^3		Remarks[2]
								kg	
									Details are shown in the attached list
5	Paint stores	Paint, xx Co., #600	20.00	kg	5	Pcs	100.00	kg	Cadmium containing

Note

[1] The location of a part II or part III item should be entered in order based on its location, from a lower level to an upper level and from a fore part to an aft part. The location of part I items is recommended to be described similarly, as far as practicable.

[2] In column "Remarks" for part III items, if hazardous materials are integrated in products, the approximate amount of the contents should be shown as far as possible.

III-2 Liquids sealed in ship's machinery and equipment

No	Type of liquids (classification in Appendix 1)	Name of machinery or equipment	Location	Approximate quantity		Remarks
1	Hydraulic oil	Deck crane hydraulic oil system	Upper Deck	15.00	m^3	
		Deck machinery hydraulic oil system	Upper deck and bosun store	200.00	m^3	
		Steering gear hydraulic oil system	Steering gear room	0.55	m^3	
2	Lubricating oil	Main engine system	Engine room	0.45	m^3	
3	Boiler water treatment	Boiler	Engine room	0.20	m^3	

III-3—Gases sealed in ship's machinery and equipment

No	Type of gases (classification in Appendix 1)	Name of machinery or equipment	Location	Approximate quantity		Remarks
1	HFC	AC system	AC room	100.00	Kg	
2	HFC	Refrigerated provision chamber machine	AC room	50.00	kg	

III-4—Regular consumable goods potentially containing hazardous material

No	Location[16]	Name of Item	Quantity	Remarks
1	Accommodation	Refrigerators	1	
2	Accommodation	Personal computers	2	

Note
[16] The location of a Part II or Part III item should be entered in order based on its location, from a lower level to an upper level and from a fore part to an aft part. The location of Part I items is recommended to be described similarly, as far as practicable

Appendix C: Example of the Development Process for Part I of the Inventory for New Ships

Objective of the typical example

This example has been developed to give guidance and to facilitate understanding of the development process for Part I of the Inventory of Hazardous Materials for new ships.

Development flow for Part I of the Inventory

Part I of the Inventory should be developed using the following three steps. However, the order of these steps is flexible and can be changed depending on the schedule of shipbuilding:

(1) Collection of hazardous materials information
(2) Utilisation of hazardous materials information; and
(3) Preparation of the Inventory (by filling out standard format).

Collections of Hazardous Materials Information

Data Collection Process for Hazardous Materials

Materials Declaration (MD) and Supplier's Declaration of Conformity (SDoC) for products from suppliers (tier 1 suppliers) should be requested and collected by the shipbuilding yard. Tier 1 suppliers may request from their suppliers (tier 2 suppliers) the relevant information if they cannot develop the Materials Declaration based on the information available. Thus, the collection of data on hazardous materials may involve the entire shipbuilding supply chain (Fig. A3.1).

© The Editor(s) (if applicable) and The Author(s), under exclusive license
to Springer Nature Switzerland AG 2025
F. Karkori and A. A. Olsen, *Inventory of Hazardous Materials*, Synthesis Lectures
on Ocean Systems Engineering, https://doi.org/10.1007/978-3-031-76734-0

Fig. A3.1 Process of materials
declaration (and Supplier's
Declaration of Conformity)
collection showing
involvement of supply chain

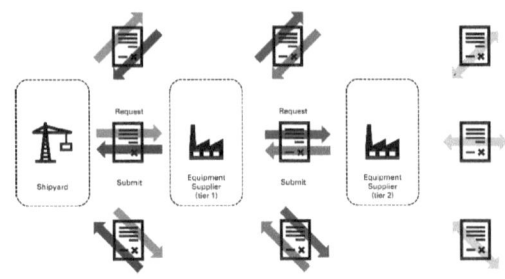

Declaration of Hazardous Materials

Suppliers should declare whether or not the hazardous materials listed in Table A and
Table B in the Material Declaration are present in concentrations above the threshold
values specified for each homogeneous material in a product.

Materials Listed in Table A

If one or more materials listed in Table A are found to be present in concentrations above
the specified threshold value according to the Material Declaration, the products which
contain these materials shall not be installed on a ship. However, if the materials are
used in a product in accordance with an exemption specified by the Convention (e.g. new
installations containing hydrochlorofluorocarbons (HCFCs) before 1 January 2020), the
product should be listed in the Inventory.

Materials Listed in Table B

If one or more materials listed in Table B are found to be present in concentrations above
the specified threshold value according to the Material Declaration, the products should
be listed in the Inventory.

Example of Homogeneous Materials

Figure A3.2 shows an example of four homogeneous materials which constitute a cable.
In this case, sheath, intervention, insulator and conductor are all individual homogeneous
materials

Utilisations of Hazardous Materials Information

Products which contain hazardous materials in concentrations above the specified threshold values should be clearly identified in the Material Declaration. The approximate
quantity of the hazardous materials should be calculated if the mass data for hazardous

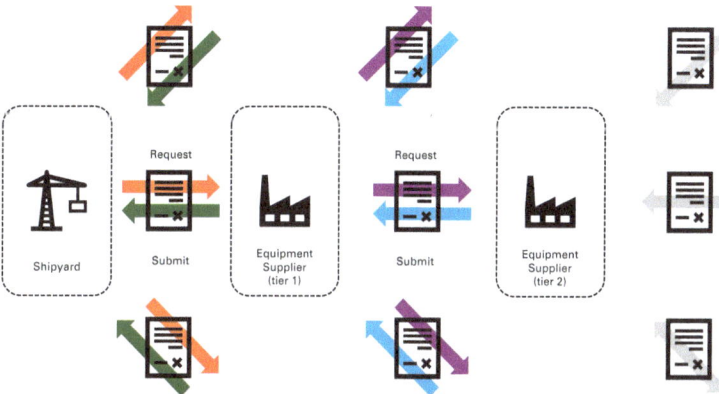

Fig. A3.2 Example of Homogeneous Materials (Cable)

materials are declared in the Material Declaration using a unit which cannot be directly utilised in the Inventory.

Preparations of Inventory (By Filling Out Standard Format)

The information received for the Inventory, as contained in Table A and Table B of Appendix 1 of this text, ought to be structured and utilised according to the following categorisation for Part I of the Inventory:

Part I-1 Paints and coating systems.

Part I-2 Equipment and machinery; and.

Part I-3 Structure and hull.

"Name of Equipment and Machinery" Column

Equipment and Machinery

The name of each item of equipment or machinery should be entered in this column. If more than one hazardous material is present in the equipment or machinery, the row relating to that equipment or machinery should be appropriately divided such that all of the hazardous materials contained in the piece of equipment or machinery are entered. If more than one item of equipment or machinery is situated in one location, both name and quantity of the equipment or machinery should be entered in the column. Examples are shown in rows 1 and 2 of Table A3.1. For identical or common items, such as but not limited to bolts, nuts and valves, there is no need to list each item individually.

Table A3.1 Example showing more than one item of equipment or machinery situated in one location

No	Name of equipment and machinery	Location	Materials (classification in Appendix 1)	Parts where used	Approx. quantity		Remarks
1	Main engine	Engine room	Lead	Piston pin bush	0.75	kg	
			Mercury	Thermometer charge air temperature	0.01	kg	
2	Diesel generator (× 3)	Engine room	Mercury	Thermometer	0.03	kg	
3	FC valve (× 100)	Throughout the ship	Lead and lead compounds		20.5	kg	

Pipes and Cables

The names of pipes and of systems, including electric cables, which are often situated in more than one compartment of a ship, should be described using the name of the system concerned. A reference to the compartments where these systems are located is not necessary as long as the system is clearly identified and properly named.

"Approximate Quantity" Column

The standard unit for approximate quantity of solid hazardous materials should be kg. If the hazardous materials are liquids or gases, the standard unit should be either m^3 or kg. An approximate quantity should be rounded up to at least two significant figures. If the hazardous material is less than 10 g, the description of the quantity should read "<0.01 kg".

Table A3.2 Example of a switchboard

No	Name of equipment and machinery	Location	Materials (classification in Appendix 1)	Parts where used	Approx. quantity		Remarks
1	Switchboard	Engine control room	Cadmium	Housing coating	0.02	kg	
			Mercury	Heat gauge	<0.01	Kg	Less than 0.01 kg

"Location" Column

Example of a Location List

It is recommended to prepare a location list which covers all compartments of a ship based on the ship's plans (e.g. general arrangement, engine-room arrangement, accommodation and tank plan) and on other documentation on board, including certificates or spare parts' lists. The description of the location should be based on a location such as a deck or room to enable easy identification. The name of the location should correspond to the ship's plans so as to ensure consistency between the Inventory and the ship's plans. Examples of names of locations are shown in Table A3.3. For bulk listings, the locations of the items or materials may be generalised. For example, the location may only include the primary classification such as "Throughout the ship" as shown in the Table A3.3.

Description of Location of Pipes and Electrical Systems

Locations of pipes and systems, including electrical systems and cables situated in more than one compartment of a ship, should be described for each system concerned. If they are situated in a number of compartments, the most practical of the following two options should be used:

(1) listing of all components in the column; or
(2) description of the location of the system using an expression such as those shown under "primary classification" and "secondary classification" in Table A3.3.

A typical description of a pipe system is shown in Table A3.4.

Table A3.3 Examples of location names

(A) Primary classification	(B) Secondary classification	(C) Name of location
Throughout the ship		
Hull part	Fore part	Bosun store
	Cargo part	No.1 cargo hold/tank
		No.1 garage deck
	Tank part	Fore peak tank
		No.1 WBT
		No.1 FOT
		Aft Peak Tank
	Aft part	Steering gear room
		Emergency fire pump space
	Superstructure	Accommodation
		Compass deck
		Navigational bridge deck
		Wheelhouse
		Engine control room
		Cargo control room
	Deck house	Deck house
Machinery part	Engine room	Engine room
		Main floor
		2nd floor
		Generator space/room
		Purifier space/room
		Shaft space/room
		Engine casing
		Funnel
		Engine control room
	Pump room	Pump room
Exterior part	Superstructure	Superstructure
	Upper deck	Upper deck
	Hull shell	Hull shell
		Bottom
		Under waterline

Table A3.4 Example of description of a pipe system

No	Name of equipment and machinery	Location	Materials (classification in Appendix 1)	Parts where used	Approx. quantity		Remarks
1	Ballast water system	Engine room, hold parts					

Appendix D: Flow Diagram for Developing Part I of the Inventory for Existing Ships

This Appendix contains a flow diagram for developing Part I of the Inventory for existing ships.

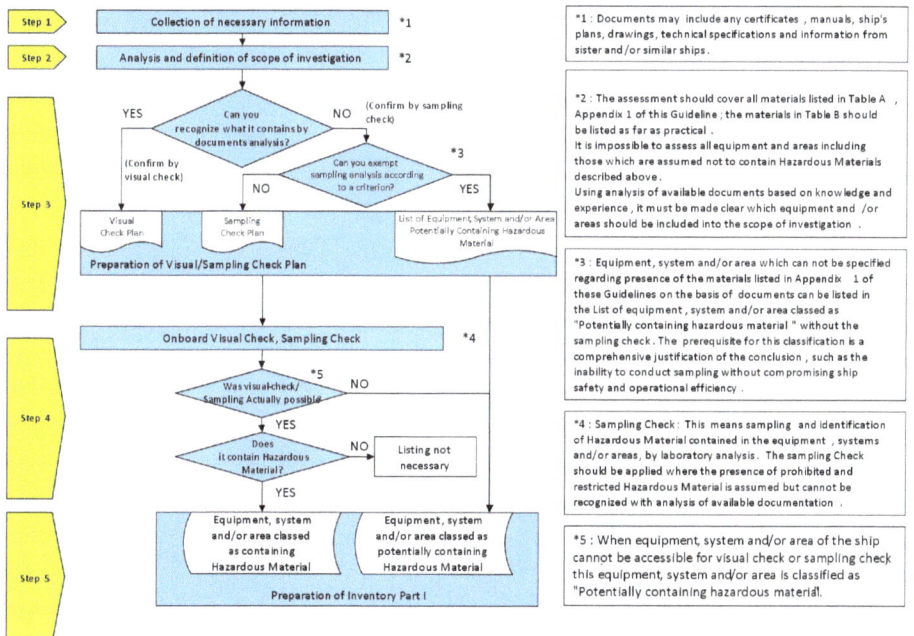

Appendix E: Example of the Development Process for Part I of the Inventory for Existing Ships

Introduction

In order to develop Part I of the Inventory of Hazardous Materials for existing ships, documents of the individual ship as well as the knowledge and experience of specialist personnel (experts) is required. An example of the development process for Part I of the Inventory of Hazardous Materials for existing ships is useful to understand the basic steps as laid out in the guidelines and to ensure a unified application. However, attention should be paid to variations in different types of ships.[1]

Compilation of Part I of the Inventory of Hazardous Material for existing ships involves the following five steps which are described in Chap. 4, Sect. 2 and Appendix 4 of this text:

Step 1 Collection of necessary information.
Step 2 Assessment of collected information.
Step 3 Preparation of visual/sampling check plan.
Step 4 Onboard visual/sampling check; and.
Step 5 Preparation of Part I of the Inventory and related documentation.

Step 1—Collection of Necessary Information

Sighting of Available Documents

A practical first step is to collect detailed documents for the ship. The shipowner should try to collate documents normally retained on board the ship or by the shipping company as well as relevant documents that the shipyard, manufacturers, or classification society may have. The following documents should be used when available:

---

[1] The example of a 28,000 gross tonnage bulk carrier constructed in 1985 is used in this appendix.

F. Karkori and A. A. Olsen, *Inventory of Hazardous Materials*, Synthesis Lectures on Ocean Systems Engineering, https://doi.org/10.1007/978-3-031-76734-0

  (1) Ship's specification
  (2) General Arrangement
  (3) Machinery Arrangement
  (4) Spare Parts and Tools List
  (5) Piping Arrangement
  (6) Accommodation Plan
  (7) Fire Control Plan
  (8) Fire Protection Plan
  (9) Insulation Plan (Hull and Machinery)
 (10) International Anti-Fouling System Certificate
 (11) Related manuals and drawings
 (12) Information from other inventories and/or sister or similar ships, machinery, equipment, materials and coatings; and
 (13) Results of previous visual/sampling checks and other analysis.

Indicative List

It is impossible to check all equipment, systems, and/or areas on board the ship to determine the presence or absence of hazardous materials. The total number of parts on board may exceed several thousand. In order to take a practical approach, an indicative list should be prepared that identifies the equipment, system, and/or area on board that is presumed to contain hazardous materials. Field interviews with the shipyard and suppliers may be necessary to prepare such lists. A typical example of an indicative list is shown below.

Materials to Be Checked and Documented
Hazardous Materials, as identified in Appendix 1, should be listed in Part I of the Inventory for existing ships. Appendix 1 contains all the materials concerned. Table A1.1 shows those which are required to be listed and Table B shows those which should be listed as far as practicable.

Materials Listed in Table A1.1
Table A1.1 lists the following four materials:

  (1) Asbestos
  (2) Polychlorinated biphenyls (PCBs)
  (3) Ozone depleting substances; and
  (4) Anti-fouling systems containing organotin compounds as a biocide.

Asbestos

Field interviews were conducted with over 200 Japanese shipyards and suppliers regarding the use of asbestos in production. Indicative lists for asbestos developed on the basis of this research are shown below:

Structure and/or equipment	Component
Propeller shafting	Packing with low pressure hydraulic piping flange
	Packing with casing
	Clutch
	Brake lining
	Synthetic stern tubes
Diesel engine	Packing with piping flange
	Lagging material for fuel pipe
	Lagging material for exhaust pipe
	Lagging material turbocharger
Turbine engine	Lagging material for casing
	Packing with flange of piping and valve for steam line, exhaust line and drain line
	Lagging material for piping and valve of steam line, exhaust line and drain line
Turbine engine	Insulation in combustion chamber
	Packing for casing door
	Lagging material for exhaust pipe
	Gasket for manhole
	Gasket for hand hole
	Gas shield packing for soot blower and other hole
	Packing with flange of piping and valve for steam line, exhaust line, fuel line and drain line
	Lagging material for piping and valve of steam line, exhaust line, fuel line and drain line
Exhaust gas economiser	Packing for casing door
	Packing with manhole
	Packing with hand hole
	Gas shield packing for soot blower
	Packing with flange of piping and valve for steam line, exhaust line, fuel line and drain line
	Lagging material for piping and valve of steam line, exhaust line, fuel line and drain line

(continued)

(continued)

Structure and/or equipment	Component
Incinerator	Packing for casing door
	Packing with manhole
	Packing with hand hole
	Lagging material for exhaust pipe
Auxiliary machinery (pump, compressor, oil purifier, crane)	Packing for casing door and valve
	Gland packing
	Brake lining
Heat exchanger	Packing with casing
	Gland packing for valve
	Lagging material and insulation
Valve	Gland packing with valve, sheet packing with piping flange
	Gasket with flange of high pressure and/or high temperature
Pipe, duct	Lagging material and insulation
Tank (fuel tank, hot water, tank, condenser), other equipment (fuel strainer, lubricant oil strainer)	Lagging material and insulation
Electric equipment	Insulation material
Airborne asbestos	Wall, ceiling
Ceiling, floor and wall in accommodation area	Ceiling, floor, wall
Fire door	Packing, construction and insulation of the fire door
Inert gas system	Packing for casing, etc
Air-conditioning system	Sheet packing, lagging material for piping and flexible joint
Miscellaneous	Ropes
	Thermal insulating materials
	Fire shields/fireproofing
	Space/duct insulation
	Electrical cable materials
	Brake linings
	Floor tiles/deck underlay
	Steam/water/vent flange gaskets
	Adhesives/mastics/fillers

(continued)

(continued)

Structure and/or equipment	Component
	Sound damping
	Moulded plastic products
	Sealing putty
	Shaft/valve packing
	Electrical bulkhead penetration packing
	Circuit breaker arc chutes
	Pipe hanger inserts
	Weld shop protectors/burn covers
	Fire-fighting blankets/clothing/equipment
	Concrete ballast

Polychlorinated Biphenyl (PCBs)

Worldwide restriction of PCBs began on 17 May 2004 as a result of the implementation of the Stockholm Convention, which aims to eliminate or restrict the production and use of persistent organic pollutants. In Japan, domestic control began in 1973, with the prohibition of all activities relating to the production, use and import of PCBs. Japanese suppliers can provide accurate information concerning their products. The indicative list of PCBs has been developed as shown below:

Equipment	Component of equipment
Transformer	Insulating oil
Condenser	Insulating oil
Fuel heater	Heating medium
Electric cable	Covering, insulating tape Lubricating
Lubricating oil	
Heat oil	Thermometers, sensors, indicators Rubber
Rubber/felt gaskets	
Rubber hose	
Plastic foam insulation	
Thermal insulating materials	
Voltage regulators	
Switches/reclosers/bushings	
Electromagnets	
Adhesives/tapes	

(continued)

(continued)

Equipment	Component of equipment
Surface contamination of machinery	
Oil-based paint	
Caulking	
Rubber isolation mounts	
Pipe hangers	
Light ballasts (component within fluorescent light fixtures)	
Plasticisers	
Felt under septum plates on top of hull bottom	

Ozone Depleting Substances

The indicative list for ozone depleting substances is shown below. Ozone depleting substances have been controlled according to the Montreal Protocol and MARPOL Convention. Although all substances have been banned since 1996, HCFC could still be used until 2020.

Materials	Component of equipment	Period for use of ODS in Japan
CFCs (R11, R12)	Refrigerant for refrigerators	Until 1996
CFCs	Urethane formed material	Until 1996
Halons	Until 1996	Until 1996
Other fully halogenated CFCs	Blowing agent for insulation of LNG carriers	Until 1996
Carbon tetrachloride	Extinguishing agent	Until 1994
1, 1, 1-Trichloroethane (methyl chloroform)	The possibility of usage in ships is low	Until 1996
HCFC (R22, R141b)	*Refrigerant for refrigerating machine*	It was possible to use until 2020
HBFC	The possibility of usage in ships is low	Until 1996
Methyl bromide	The possibility of usage in ships is low	Until 2005

Organotin Compounds

Organotin compounds include tributyl tins (TBT), triphenyl tins (TPT) and tributyl tin oxide (TBTO). Organotin compounds have been used as anti-fouling paint on ships' bottoms and the International Convention on the Control of Harmful Anti-Fouling Systems on Ships (AFS Convention) stipulates that all ships shall not apply or re-apply organotin compounds after 1 January 2003, and that, after 1 January 2008, all ships shall either not bear such compounds on their hulls or shall bear a coating that forms a barrier preventing such compounds from leaching into the sea. The above-mentioned dates may have been extended by permission of the Administration bearing in mind that the AFS Convention entered into force on 17 September 2008.

Materials Listed in Table B

For existing ships, it is not obligatory for materials listed in Table A1.2 to be listed in Part I of the Inventory. However, if they can be identified in a practical way, they should be listed in the Inventory, because the information will be used to support ship recycling processes. The Indicative list of materials listed in Table A1.2 is shown below:

Materials	Component of equipment
Cadmium and cadmium compounds	Plating film, bearing
Hexavalent chromium compounds	Plating film
Mercury and mercury compounds	Fluorescent light, mercury lamp, mercury cell, liquid-level switch, gyro compass, thermometer, measuring tool, manganese cell, pressure sensors, light fittings, electrical switches, fire detectors
Lead and lead compounds	Corrosion resistant primer, solder (all electric appliances contain solder), paints, preservative coatings, cable insulation, lead ballast, generators
Polybrominated biphenyls (PBBs)	Non-flammable plastics
Polybrominated diphenyl ethers (PBDE)	Non-flammable plastics
Polychlorinated naphthalene	Paint, lubricating oil
Radioactive substances	Refer to Appendix 10
Certain short chain chlorinated paraffins	Non-flammable plastics

Step 2—Assessment of Collected Information

Preparation of a checklist is an efficient method for developing the Inventory for existing ships in order to clarify the results of each step. Based on collected information including the indicative list mentioned in Step 1, all equipment, systems, and/or areas on board assumed to contain hazardous materials listed in Tables A and B should be included in the checklist. Each listed equipment, system, and/or area on board should be analysed and assessed for its hazardous materials content. The existence and volume of hazardous materials may be judged and calculated from the Spare parts and tools list and the maker's drawings. The existence of asbestos contained in floors, ceilings and walls may be identified from Fire Protection Plans, while the existence of TBT in coatings can be identified from the International Anti-Fouling System Certificate, Coating scheme and the History of Paint.

Example of Weight Calculation

No	Hazardous Materials	Location/equipment/component	Reference	Calculation
1.1–2	TBT	Flat bottom/paint	History of coatings	
1.2–1	Asbestos	Main engine/exhaust pipe packing	Spare parts and tools list	250 g × 14 sheet = 3.50 kg
1.2–3	HCFC	Ref. provision plant	Maker's drawings	20 kg × 1 cylinder = 20 kg
1.2–4	Lead Batteries	Maker's drawings	6 kg × 16 unit = 96 kg	
1.3–1	Asbestos	Engine-room ceiling	Accommodation plan	

When a component or coating is determined to contain hazardous materials, a "Y" should be entered in the column for "Result of document analysis" in the checklist, to denote "Contained." Likewise, when an item is determined not to contain Hazardous Materials, the entry "N" should be made in the column to denote "Not contained." When a determination cannot be made as to the hazardous materials content, the column should be completed with the entry "Unknown."

Check List (Step 2) Analysis and Definition of Scope of Assessment for "Sample Ship"

No	Table # A/B	Hazardous Materials #1	Location	Name of Equipment	Component	Unit (Kg)	No	Total (Kg)	Manufacturer/ Brand Name	Result of Document analysis #2	Procedure of check #3	Results of check #4	Reference / Dwg. No
[Inventory Part 1.1.1]													
1	A	TBI	Top side	Painting and coating	A/F Paints			Nil	Paints Co/ Maiine P1000	N			On Aug. 200X sea er coat applied to all over submerged area before tin free coating
2	A	TBI	Flat bottom				3000 m3		Unknown AF	Unknown			
[Inventory Part 1.1.2]													
1	A	Asbestos	Lower deck	Main engine	Exhaust pipe packing	0.25	14		Diesel Co	Y			M100
2	A	Asbestos	3rd deck	Aux. boiler	Lagging		12		Unknown lagging	Unknown			M30Q
3	A	Asbestos	Engine room	Piping/flange	Packing					PCHM			
4	A	HCFC	2nd deck	Ref. provision plant	Refrigerant R22	20.0	1		Reito Co	Y			Makers Dwg
5	B	Lead	Nav. Br. deck	Batteries		6	16		Denchi Co	Y			E-300
[Inventory Part 1.1.3]													
1	A	Asbestos	Upper deck	Back deck ceilings	Engine room ceilings		20 m3		Unknown ceiling	Unknown			0-25

Note

(1) *Hazardous materials: materials classification*
(2) *Result of documents analysis: Y = contained, N = Not contained, Unknown, PCHM = Potentially containing hazardous material*
(3) *Procedure of check: V = visual check, S = Sampling check*
(4) *Result of check: Y = Contained, N = Not Contained, PCHM = Potentially containing hazardous material*

Step 3—Preparation of visual/Sampling Check Plan+

Each item classified as "Contained" or "Not contained" in step 2 should be subjected to a visual check on board, and the entry "V" should be made in the "Check procedure" column to denote "Visual check." For each item categorised as "unknown," a decision should be made as to whether to apply a sampling check. However, any item categorised as "unknown" may be classed as "potentially containing hazardous material" provided comprehensive justification is given, or if it can be assumed that there will be little or no effect on disassembly as a unit and later ship recycling and disposal operations. For example, in the following checklist, in order to carry out a sampling check for "Packing with aux. boiler" the shipowner needs to disassemble the auxiliary boiler in a repair yard. The costs of this check are significantly higher than the later disposal costs at a ship recycling facility. In this case, therefore, the classification as "potentially containing hazardous material" is justifiable.

Check List (Step 3) Analysis and Definition of Scope of Assessment for "Sample Ship"

No	Table # A/B	Hazardous Materials #1	Location	Name of Equipment	Component	Quantity of Hazmat Unit (Kg)	No	Total (Kg)	Manufacturer' Brand Name	Result of Document analysis #2	Procedure of check #3	Results of check #4	Reference 1 Dwg, Mo
[Inventory Part 1.1.1]													
1	A	TBT	Top side	Painting and coating	A/F Paints			Nil	Paints Co/Marine P1000	N	V		On Aug. 200X sealer coat applied to all over submerged area before tin free coating
2	A	TBT	Flat bottom				3000 m³		Unknown AF	Unknown	S		
[Inventory Part 1.1.2]													
1	A	Asbestos	Lower deck	Main engine	Exhaust pipe packing	0.25	14		Diesel Co	Y	V		M100
2	A	Asbestos	3rd deck	Aux. boiler	Lagging		12		Unknown lagging	Unknown	s		M300
3	A	Asbestos	Engine room	Piping/flange	Packing					PCHM	V		
4	A	HCFC	2nd deck	Ref. provision plant	Refrigerant R22	20.0	1		Reito Co	Y	V		Makers Dwg
5	B	Lead	Nav. Br. deck	Batteries		6	16		Denchi Co	Y	V		E-300
[Inventory Part 1.1.3]													
1	A	Asbestos	Upper deck	Back deck ceilings	Engine room ceilings		20 m²		Unknown ceiling	Unknown	s		0–25

Note

(1) *Hazardous materials: materials classification*
(2) *Result of documents analysis: Y = contained, N = Not contained, Unknown, PCHM = Potentially containing hazardous material*
(3) *Procedure of check: V = visual check, S = Sampling check*
(4) *Result of check: Y = Contained, N = Not Contained, PCHM = Potentially containing hazardous material*

Before any visual/sampling check on board is conducted, a "visual/sampling check plan" should be prepared. An example of such a plan is shown below. To prevent any incidents during the visual/sampling check, a schedule should be established to eliminate interference with other ongoing work on board. To prevent potential exposure to Hazardous Materials during the visual/sampling check, safety precautions should be in place on board. For example, sampling of potential asbestos containing materials could release fibres into the atmosphere. Therefore, appropriate personnel safety and containment procedures should be implemented prior to sampling. Items listed in the visual/sampling check should be arranged in sequence so that the onboard check is conducted in a structured manner (e.g. from a lower level to an upper level and from a fore part to an aft part).

Example of Visual/sampling Check Plan

Name of ship	*XXXXXXXXXX*
IMO Number	*XXXXXXXXXX*
Gross Tonnage	*28,000 GT*
L x B x D	*xxx.xx × xx.xx × xx.xx m*
Date of delivery	*dd.mm.1987*
Shipowner	*XXXXXXXXXX*
Contact point (Address, telephone, fax, email)	*Tel: XXXX-XXXX* *Fax:: XXXX-XXXX* *Email: abcdefg@abc.co.uk*
Check schedule	*Visual check: dd, mm, 20XX* *Sampling check: dd, mm, 20XX*
Site of check	*XXXX XXXX*
In charge of check	*XXXX XXXX, YYYY YYYY, ZZZZ ZZZZ*
Sampling engineer	*Person with specialised knowledge of sampling*
Sampling method	*Wet the sampling location prior to cutting and allow it to harden after cutting to prevent scatter* ***Note:*** *Workers performing sampling activities shall wear protective equipment*

(continued)

(continued)

Name of ship	*XXXXXXXXXX*
Sampling of fragments of paints	*Paints suspected to contain TBT should be collected and analysed from load line, directly under bilge keel and flat bottom near amidships*
Laboratory	*QQQQ QQQQ*
Chemical analysis method	*Method by ISO/DIS 22262–1 Bulk materials—Part 1: Sampling and qualitative determination of asbestos in commercial bulk materials and ISO/CD 22,262–2 Bulk materials—Part 2: Quantitative determination of asbestos by gravimetric and microscopic methods. ICP Luminous analysis (TBT)*
Location of visual/sampling check	*Refer to lists for visual/sampling check*

List of equipment, system and/or area for sampling check

Location	Equipment, machinery and/or zone	Name of parts	Materials	Result of doc. checking
Upper Deck	Back deck ceilings	Engine-room ceiling	Asbestos	Unknown
Engine-room	Exhaust gas pipe	Insulation	Asbestos	Unknown
Engine-room	Pipe/flange	Gasket	Asbestos	Unknown

List of equipment, system and/or area classed as PCHM

Location	Equipment, machinery and/or zone	Name of part	Material	Result of doc. checking
Floor	Propeller cap	Gasket	Asbestos	PCHM
Engine-room	Air operated shut-off valve	Gland packing	Asbestos	PCHM

Refer to attached "Analysis and definition of scope of investigation for sample ship " and "Location plan of hazardous materials for sample ship"

This plan is established in accordance with the guidelines for the development of the Inventory of Hazardous Materials

Prepared by: XXXX XXXX.

Tel.: YYYY-YYYY.

Email: XXXX@ZZZZ.co.net.

Document check • date/place

dd, mm, 20XX at XX Lines Co. Ltd.

Preparation date of plan: dd. mm, 20XX.

Step 4—Onboard Visual/Sampling Check

The visual/sampling check should be conducted according to the plan. Check points should be marked in the ship's plan or recorded with photographs. A person taking samples should be protected by the appropriate safety equipment relevant to the suspected type of hazardous materials encountered. Appropriate safety precautions should also be in place for passengers, crewmembers and other persons on board, to minimise the potential exposure to hazardous materials. Safety precautions could include the posting of signs or other verbal or written notification for personnel to avoid such areas during sampling. The personnel taking samples should ensure compliance with relevant national regulations. The results of visual/sampling checks should be recorded in the checklist. Any equipment, systems and/or areas of the ship that cannot be accessed for checks should be classified as "potentially containing hazardous material." In this case, the entry in the "Result of check" column should be "PCHM."

Step 5—Preparation of Part I of The Inventory and Related Documentation

Development of Part I of the Inventory

The results of the check and the estimated quantity of hazardous materials should be recorded on the checklist. Part I of the Inventory should be developed with reference to the checklist.

Development of Location Diagram of Hazardous Materials

With respect to Part I of the Inventory, the development of a location diagram of hazardous materials is recommended in order to help the ship recycling facility gain a visual understanding of the Inventory.

Check List (Step 4 and Step 5) Analysis and Definition of Scope of Assessment for "Sample Ship"

No	Table # A/B	Hazardous Materials #1	Location	Name of Equipment	Component	Quantity of Hazmat Unit (Kg)	No	Total (Kg)	Manufacturer/ Brand Nam©	Result of Document analysis #2	Procedure of check #3	Results of check #4	Reference 1 Dwg. No
[Inventory Part 1.1.1]													
1	A	TBT	Top side	Painting and coating	A/F Paints			Nil	Paints Co/ Marine P1000	V	V	N	On Aug. 200X sealer coat applied to all over submerged area before tin free coating
2	A	TBT	Flat bottom			.02	3000 m^3	00.00	Unknown AF	S	S	Y	
[Inventory Part 1.1.2]													
1	A	Asbestos	Lower deck	Main engine	Exhaust pipe packing	0.25	14	3.50	Diesel Co	Y	V	Y	M100
2	A	Asbestos	3rd deck	Auk. boiler	Lagging		12		Unknown lagging	Unknown	S	N	M300
3	A	Asbestos	Engine room	Piping/flange	Packing					PCHM	V	PCHM	
4	A	HCFC	2nd deck	Ref. provision plant	Refrigerant R22	20.0	1	20.0	Reito Co	Y	V	Y	Makers Dwg
5	B	Lead	Nav. Br. deck	Batteries		6	16	96.00	Denchii Co	Y	V	Y	E-300
[Inventory Part 1.1.3]													
1	A	Asbestos	Upper deck	Back deck ceilings	Engine room ceilings	0.19	20 m^2	3.80	Unknown ceiling	Unknown	s	Y	0–25

Note

(1) *Hazardous materials: materials classification*
(2) *Result of documents analysis: Y = contained, N = Not contained, Unknown, PCHM = Potentially containing hazardous material*
(3) *Procedure of check: V = visual check, S = Sampling check*
(4) *Result of check: Y = Contained, N = Not Contained, PCHM = Potentially containing hazardous material*

Example of the Inventory for existing ships Inventory of Hazardous Materials for "Sample Ship" Particulars of the "Sample Ship"

Distinctive number or letters	Distinctive number or letters
XXXXNNN	XXXXNNN
Port of registry	Port of registry
Port of World	Port of World
Type of vessel	Type of vessel
Bulk carrier	Bulk carrier
Gross Tonnage	Gross Tonnage
28,000 GT	28,000 GT

This inventory was developed in accordance with the guidelines for the development of the Inventory of Hazardous Materials.

Attachment:

(1) Inventory of Hazardous Materials
(2) Assessment of collected information; and
(3) Location diagram of Hazardous Material.

Prepared by XYZ (Name & address) (dd/mm/20XX).

Inventory of Hazardous Materials: "Sample Ship"
Part I—Hazardous materials contained in the ship's structure and equipment.

Example of Location diagram of Hazardous Materials

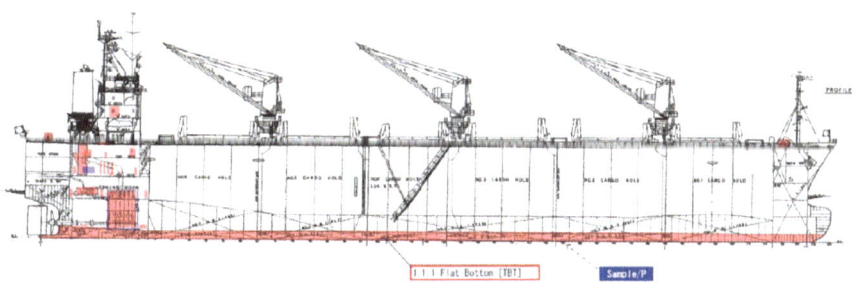

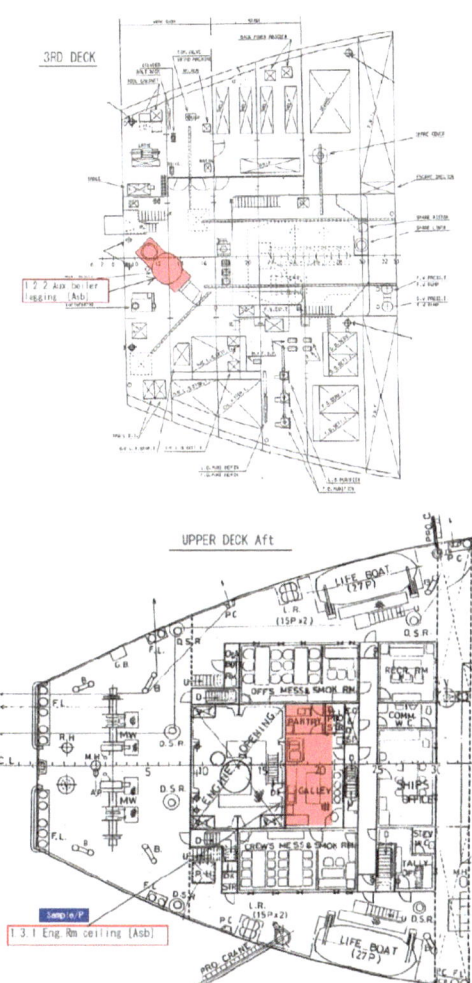

Appendix F: Form of Material Declaration

<Date of declaration>

Date	

<MD ID number>

MD-ID –No.	

<Other information>

Remark 1	
Remark 2	
Remark 3	

<Supplier (respondent) information>

Company name	
Division name	
Address	
Contact person	
Telephone number	
Fax number	
E-mail Address	
SDoC ID no.:	

<Product information>

Product name	Product number	Delivered Unit			Product information
		Amount	Unit		

<Materials information>

This materials information shows the amount of hazardous materials contained in

	Unit	
1	

(unit: piece. kg, m, m2, m3, etc.) of the product

Table	Material name		Threshold value	Present above threshold value	If yes, material mass		If yes, information on where it is used
				Yes / No	Mass	Unit	
Table A (materials listed in appendix 1 of the Convention)	Asbestos	Asbestos	0.1% [18]				
	Polychlorinated biphenyls (PCBs)	Polychlorinated biphenyls (PCBs)	50 mg/kg				
	Ozone depleting substances	Chlorofluorocaobans (CFCs)	no threshold value				
		Halons					
		Other fully halogenated CFCs					
		Carbon tetrachloride					
		1,1,1-Trichloroethane					
		Hydrochlorofluorocaobons					
		Hudrobromofluorocaobons					
		Methyl bromide					
		Bromochloromethane					

Note

F. Karkori and A. A. Olsen, *Inventory of Hazardous Materials*, Synthesis Lectures
on Ocean Systems Engineering, https://doi.org/10.1007/978-3-031-76734-0

[18]In accordance with regulation 4 of the Convention, for all ships, new installation of materials which contain asbestos shall be prohibited. According to the UN recommendation "Globally Harmonised System of Classification and Labelling of Chemicals (GHS)" adopted by the United Nations Economic and Social Council's Sub Committee of Experts on the Globally Harmonised System of Classification and Labelling of Chemicals (UNSCEGHS), the UN's Sub-Committee of Experts, in 2002 (published in 2003), carcinogenic mixtures classified as Category 1A (including asbestos mixtures) under the GHS are required to be labelled as carcinogenic if the ratio is more than 0.1%. However, if 1% is applied, this threshold value should be recorded in the Inventory and, if available, the Material Declaration and can be applied not later than five years after the entry into force of the Convention. The threshold value of 0.1% need not be retroactively applied to those Inventories and Material Declarations.

Table	Material name	Threshold value	Present above threshold value	If yes, material mass		If yes, information on where it is used
			Yes / No	Mass	Unit	
Table A (continued)	Anti-fouling systems containing organotin compounds as a biocide	2,500 mg total tin/kg				
Table B (materials listed in appendix 2 of the Convention)	Cadmium and cadmium compounds	100 mg/kg				
	Hexavalent chromium and hexavalent chromium compounds	1,000 mg/kg				
	Lead and lead compounds	1,000 mg/kg				
	Mercury and mercury compounds	1,000 mg/kg				
	Polybrominated biphenyl (PBBs)	50 mg/kg				
	Polybrominated diphenyl ethers (PBDEs)	1,000 mg/kg				
	Polychloronaphalenes (Cl > = 3)	50 mg/kg				
	Radioactive substances	No threshold value				

(continued)

(continued)

Table	Material name	Threshold value	Present above threshold value	If yes, material mass		If yes, information on where it is used
			Yes / No	Mass	Unit	
	Certain shortchain chlorinated paraffins	1%				

Appendix G: Form of Supplier's Declaration of Conformity

1) Identification number: _____

2) Issuer's name: _____

    Issuer's address: _____

3) Object(s) of the declaration: _____

    _____

    _____

4) The object(s) of the declaration described above is in conformity with the following documents:

    Document no.   Title   Edition/date of issue

5) _____   _____   _____

    _____   _____   _____

    _____   _____   _____

6) Additional information: _____

    _____

    Signed for and on behalf of:

    _____

    _____

    (place and date of issue)

7) _____   _____

    (name, function)   (signature)

Appendix H: Examples of Table A and Table B Materials of Appendix 1 with Cas

Numbers

This list was developed with reference to Joint Industry Guide No.101. This list is not exhaustive; it represents examples of chemicals with known CAS* numbers and may require periodical updating.

Table	Material Category	Substances	CAS Numbers
		Asbestos	1332–21-4
		Actinolite	77,536–66-4
		Amosite (Grunerite)	12,172–73-5
	Asbestos	Anthophyllite	77,536–67-5
		Chrysotile	12,001–29-5
		Crocidolite	12,001–28-4
		Tremolite	77,536–68-6
		Polychlorinated biphenyls	1336–36-3
	Polychlorinated biphenyls (PCBs)	Aroclor	12,767–79-2
		Chlorodiphenyl (Aroclor 1260)	11,096–82-5
		Kanechlor 500	27,323–18-8
		Aroclor 1254	11,097–69-1
		Trichlorofluoromethane (CFC11)	75–69-4
		Dichlorodifluoromethane (CFC12)	75–71-8
		Chlorotrifluoromethane (C'FC 13)	75–72-9
		Pentachlorofluoroethane (C'FC 111)	354–56-3
		Tetrachlorodifluoroethane (CFC 112)	76–12-0
		Trichlorotrifluoroethane (CFC 113)	354–58-5
		1,1,2 Trichloro-1,2,2 trifluoroethane	76–13-1
		Dichlorotetrafluoroethane (CFC 114)	76–14-2

(continued)

F. Karkori and A. A. Olsen, *Inventory of Hazardous Materials*, Synthesis Lectures on Ocean Systems Engineering, https://doi.org/10.1007/978-3-031-76734-0

(continued)

Table	Material Category	Substances	CAS Numbers
		Monochloropentafluoroethane (CFC 115)	76–15-3
		Heptachlorofluoropropane (CFC 211)	422–78-6
			135,401–87-5
Table A (materials		Hexachlorodifluoropropane (CFC 212)	3182–26-1
listed in appendix 1		Pentachlorotrifluoropropane (CFC 213)	2354–06-5
of the Convention)			134,237–31-3
		Tetrachlorotetrafluoropropane (CFC 214) 1,1,1,3-	29,255–31-0
		Tetrachlorotetrafluoropropane	2268–46-4
	Ozone depleting	Trichloropentafluoropropane (CFC 215) 1.1.1-	1599–41-3
	substances/isomers	Trichloropentafluoropropane	4259–43-2
	(they may contain	1,2,3-Trichloropentafluoropropane	76–17-5
	isomers that are not	Dichlorohexafluoropropane (CFC 216)	661–97-2
	listed here)	Monochloroheptafluoropropane (CFC 217)	422–86-6
		Bromochlorodifluoromethane (Halon 1211)	353–59-3
		Bromotrifluoromethane (Halon 1301)	75–63-8
		Dibromotetrafluoroethane (Halon 2402)	124–73-2
		Carbon tetrachloride (Tetrachloromethane)	56–23-5
		1.1.1.—Trichloroethane (methyl chlorofonn) and its isomers except 1.1.2-trichloroethane	71–55-6
		Bromomethane (Methyl bromide)	74–83-9
		Bromodifluoromethane and isomers (HBFC's)	1511–62-2
		Dichlorofluoromethane (HCFC 21)	75–43-4
		Chlorodifluoromethane (HCFC 22)	75–45-6
		Chlorofluoromethane (HCFC 31)	593–70-4

(continued)

(continued)

Table	Material Category	Substances	CAS Numbers
		Tetrachlorofluoroethane (121) HCFC	134,237–32-4
		1,1,1-tetrachloro-2-fluoroethane (HCFC 121a) 1,1,2,2-tetracloro-1 -fluoroethane	354–11-0 354-14-3
		Trichlorodifluoroethane (HCFC 122) 1,2.2-trichloro-1.1 -difluoroethane	41,834–16-6 35,421–2

Table	Material Category	Substances	CAS Numbers
Table A (materials listed in appendix 1 of the Convention) (continued)	Ozone depleting substances/ isomers (they may contain isomers that are not listed here) (continued)		34,077–87-7 90,454–18-5 306–83-2 354–23-4 812–04-4 812–04-04
		Chlorotetrafluoroethane (HCFC 124) 2-chloro-1,1,1,2-tetrafluoroethane 1-chloro-1,1,2,2-tetrafluoroethane (HCFC 124a)	63,938–10-3–2837-89–0 354–25-6
		Trichlorofluoroethane (HCFC 131) 1-Fluoro-1,2,2-trichloroethane 1,1,1-trichloro-2-fluoroethane (HCFC131b)	27,154–33-2; & (134,237–34-6) & 359–28-4 811–95-0
		Dichlorodifluoroethane (HCFC 132) 1,2-dichloro-1,1-difluoroethane (HCFC 132b) 1,1-dichloro-1,2-difluoroethane (HFCF 132c) 1,1-dichloro-2,2-difluoroethane 1,2-dichloro-1,2-difluoroethane	25,915–78-0 1649 08–71,842-05–3 471–43-2 431–06-1
		Chlorotrifluoroethane (HCFC 133) 1-chloro-1,2,2-trifluoroethane 2-chloro-1,1,1-trifluoroethane (HCFC-133a)	1330–45-61,330- 45–675-88–7
		Dichlorofluoroethane(HCFC 141) 1,1-dichloro-1-fluoroethane (HCFC-141b) 1,2-dichloro-1-fluoroethane	1717–00-6; (25,167–88-8) 1717–00-6 430–57-9
		Chlorodifluoroethane (HCFC 142) 1-chloro-1,1-difluoroethane (HCFC142b) 1-chloro-1,2-difluoroethane (HCFC142a)	25,497–29-4 75–68-3 25,497–29-4
		Hexachlorofluoropropane (HCFC 221)	134,237–35-7

(continued)

(continued)

Table	Material Category	Substances	CAS Numbers
		Pentachlorodifluoropropane (HCFC 222)	134,237–36-8
		Tetrachlorotrifluropropane (HCFC 223)	134,237–36-9
		Trichlorotetrafluoropropane (HCFC 224)	134,237–38-0
		Dichloropentafluoropropane, (Ethyne, fluoro-) (HCFC 225) 2,2-Dichloro-1,1,1,3,3-pentafluoropropane (HCFC 225aa) 2,3-Dichloro-1,1,1,2,3-pentafluoropropane (HCFC 225ba) 1,2-Dichloro-1,1,2,3,3-pentafluoropropane (HCFC 225bb) 3,3-Dichloro-1,1,1,2,2-pentafluoropropane (HCFC 225ca) 1,3-Dichloro-1,1,2,2,3-pentafluoropropane (HCFC 225cb) 1,1-Dichloro-1,2,2,3,3-pentafluoropropane (HCFC 225 cc) 1,2-Dichloro-1,1,3,3,3-pentafluoropropane (HCFC 225da) 1,3-Dichloro-1,1,2,3,3-pentafluoropropane (HCFC 225ea) 1,1-Dichloro-1,2,3,3,3-pentafluoropropane (HCFC 225eb)	127,564–92-5; (2713–09-9) 128,903–21-9 422–48-0 422–44-6 422–56-0 507–55-1 13,474–88-9 431–86-7 136,013–79-1 111,512–56-2
		Chlorohexafluoropropane (HCFC 226)	134,308–72-8
		Pentachlorofluoropropane (HCFC 231)	134,190–48-0
		Tetrachlorodifluoropropane (HCFC 232)	134,237–39-1
		Trichlorotrifluoropropane (HCFC 233) 1,1,1-Trichloro-3,3,3-trifluoropropane	134,237–40-7 7125–83-9
		Dichlorotetrafluoropropane (HCFC 234)	127,564–83-4
		Chloropentafluoropropane (HCFC 235) 1-Chloro-1,1,3,3,3-pentafluoropropane	134,237–41-5 460–92-4
		Tetrachlorofluoropropane (HCFC 241)	134,190–49-1
		Trichlorodifluoropropane (HCFC 242)	134,237–42-6
		Dichlorotrifluoropropane (HCFC 243) 1,1-dichloro-1,2,2-trifluoropropane 2,3-dichloro-1,1,1-trifluoropropane 3,3-Dichloro-1,1,1-trifluoropropane	134,237–43-7 7125–99-7 338–75-0 460–69-5
		Chlorotetrafluoropropane (HCFC 244) 3-chloro-1,1,2,2-tetrafluoropropane	134,190–504- 679–85-6

(continued)

(continued)

Table	Material Category	Substances	CAS Numbers
		Trichlorofluoropropane (HCFC 251) 1,1,3-trichloro-1-fluoropropane	134,190–51-5 819–99-5
		Dichlorodifluoropropane (HCFC 252)	134,190–52-6

Table	Material Category	Substances	CAS Numbers
	Ozone depleting	Chlorotrifluoropropane (HCFC 253) 3-chloro-**1,1,1** -trifluoropropane (HCFC 253th)	134,237–44-8 460–35-5
	substances/ isomers (they may contain	Dichlorofluoropropane (HCFC 261) 1.1 -dichloro-1 -fluoropropane	134,237–45-9 7799–56-6
	isomers that are not listed here)	Chlorodifluoropropane (HCFC 262) 2-chloro-1.3-difluoropropane	134,190–53-7 102,738–79-4
	(continued)	Chlorofluoropropane (HCFC 271) 2-chloro-2-fluoropropaue	134,190–54-8 420–44-0
		Bis(tri-n-butyltiii) oxide	56–35-9
		Triphenyltin N.N'-dimethyldithiocarbainate	1803–12-9
		Triphenyltin fluoride	379–52-2
		Triphenyltin acetate	900–95-8
		Triphenyltin chloride	639–58-7
		Triphenyltin hydroxide	76–87-9
		Triphenyltin fatty acid salts (C = 9-l 1)	47,672–31-1
Table A (materials		Triphenyltin chloroacetate	7094–94-2
listed in appendix 1		Tributyltin methaeiylate	2155–70-6
of the Convention)		Bis(tributyltin) fuiuarate	6454–35-9
(continued)		Tributyltin fluoride	1983–10-4
	Oreanotin compounds (tributyl tin, triphenvl tin. tributyl tin oxide)	Bis(tiibutyltin) 2,3-dibromosuccinate	31,732–71-5
		Tributyltin acetate	56–36-0
		Tributyltin laurate	3090–36-6
		Bis(tributyltin) phthalate	4782–29-0

(continued)

(continued)

Table	Material Category	Substances	CAS Numbers
		Copolymer of alkyl aciylate, methyl methaeiylate and tributyltin methaciylate(alkyl; C = 8)	-
		Tributyltin sulfamate	6517–25-5
		Bis(tributyltin) maleate	14,275–57-1
		Tributyltin chloride	1461–22-9
		Mixture of tributyltin cyclopentanecarboxylate and its analogs (Tributyltin naphthenate)	-
		Mixture of tributyltin 1,2,3.4,4a, 4b, 5,6.10.10adecahydro-7- isopropyl-1. 4a-diinethyl-l-phenanthlenecarboxylate and its analogs (Tributyltin rosin salt)	-
		Other tributyl tins & triphenyl tins	-
		Cadmium	7440–43-9
		Cadmium oxide	1306–19-0
	Cadmium'	Cadmium sulfide	1306–23-6
	cadmium compounds	Cadmium chloride	10,108–64-2
		Cadmium sulfate	10,124–36-4
		Other cadmium compounds	-
		Chromium (VI) oxide	1333–82-0
		Barium chromate	10,294–40-3
		Calcium chromate	13,765–19-0
		Chromium trioxide	1333–82-0
		Lead (II) chromate	7758–97-6
	Chromium VT	Sodium chromate	7775–11-3
Table B (Materials listed in appendix 2 of the Convention)	compounds	Sodium dichromate	10,588–01-9
		Strontium chromate	7789–06-2
		Potassium dichromate	7778–50-9
		Potassium chromate	7789–00-6
		Zinc chromate	13,530–65-9
		Other hexavalent chromium compounds	-
		Lead	7439–92-1
		Lead (II) sulfate	7446–14-2

(continued)

(continued)

Table	Material Category	Substances	CAS Numbers
		Lead (II) carbonate	598–63-0
		Lead hydrocarbonate	1319–46-6
	Lead'lead compounds	Lead acetate	301–04-2
		Lead (II) acetate, trihydrate	6080–56-4
		Lead phosphate	7446–27-7
		Lead selenide	12,069–00-0
		Lead (IV) oxide	1309–60-0
		Lead (II,IV) oxide	1314–41-6

Table	Material Category	Substances	CAS Numbers
		Lead (II) sulfide	1314–87-0
		Lead (II) oxide	1317–36-8
		Lead (II) carbonate basic	1319–46-6
		Lead hydroxidcarbonate	1344–36-1
	Lead'lead compounds (continued)	Lead (II) phosphate	7446–27-7
		Lead (II) chromate	7758–97-6
		Lead (II) titanate	12,060–00-3
		Lead sulfate, sulphuric acid, lead salt	15,739–80-7
		Lead sulphate, tribasic	12,202–17-4
		Lead stearate	1072–35-1
		Other lead compounds	-
		Mercury	7439–97-6
		Mercuric chloride	33,631–63-9
		Mercury (**II**) chloride	7487–94-7
	Mercury/mercury compounds	Mercuric sulfate	7783–35-9
		Mercuric nitrate	10,045–94-0
		Mercuric (II) oxide	21,908–53-2
		Mercuric sulfide	1344–48-5
		Other mercury compounds	-
			2052–07-5
			(2 -Bromobiphenyl)

(continued)

(continued)

Table	Material Category	Substances	CAS Numbers
			2113–57-7
		Bromobiphenvl and its ethers	(3-Bromobiphenyl)
			92–66-0
Table B (Materials listed in appendix 2 of the Convention) (continued)			(4-Bromobiphenyl)
			101–55-3 (ether)
	Polybrominated biphenyls (PBBs) and polybrominated diphenyl ethers (PBDEs)	Dibromobiphenyl and its ethers	92–86-4
			2050–47-7 (ether)
		Nonabromobiphenvlether	63,936–56-1
		Octabromobiphenyl and its ethers	61,288–13-9
			32,536–52-0 (ether)
		Pentabromobidphenyl ether (note: commercially available PeBDPO is a complex reaction mixture containing a variety of brominated diphenyloxides	32,534–81-9 (CAS number used for commercial grades of PeBDPO)"
		Polybrominated biphenyls	59,536–65-1
		Tetrabromobiphenyl and its ethers	40,088–45-7
			40,088–47-9 (ether)
		Tribromobiphenyl ether	49,690–94-0
	Polychlorinated	Polychlorinated naphthalenes	70,776–03-3
	naphthalenes	Other polychlorinated naphthalenes	-
		Uranium	-
		Plutonium	-
		Radon	-
	Radioactive substances	Americium	-
		Thorium	-
		Cesium	7440–46-2
		Strontium	7440–24-6
		Other radioactive substances	-

(continued)

(continued)

Table	Material Category	Substances	CAS Numbers
	Certain shortchain	Chlorinated paraffins (C10-13)	85,535–84-8
	chlorinated paraffins (with carbon length of 10–13 atoms)	Other short chain chlorinated paraffins	-

Note

A CAS Number is a short string of text that refers to a chemical substance. CAS Numbers contain a sequence of up to ten numerical digits separated into three groups by two hyphens.

Appendix I: Specific Test Methods

Asbestos

Types to test for: as per resolution MEPC.179[59]; Actinolite CAS 77536–66-4 Amosite (Grunerite) CAS 12172–73-5 Anthophyllite CAS 77536–67-5 Chrysotile CAS 12001–29-5 Crocidolite CAS 12001–28-4 Asbestos Tremolite CAS 77536–68-6. Specific testing techniques: Polarised Light Microscopy (PLM), electron microscope techniques and/or Xray Diffraction (XRD) as applicable.

Specific reporting information: The presence/no presence of asbestos, indicate the concentration range, and state the type when necessary.

Notes

(1) The suggested three kinds of testing techniques are most commonly used methods when analysing asbestos and each of them has its limitation. Laboratories should choose the most suitable methods to determine, and in most cases, two or more techniques should be utilised together.

(2) The quantification of asbestos is difficult at this stage, although the XRD technique is applicable. Only a few laboratories conduct the quantification rather than the qualification, especially when a precise number is required. Considering the demand from the operators and ship recycling parties, the precise concentration is not strictly required. Thereby, the concentration range is recommended to report, and the recommended range division according to standard VDI 3866 is as follows:

- *Asbestos not detected*
- *Traces of asbestos detected*
- *Asbestos content approx. 1% to 15% by mass*
- *Asbestos content approx. 15% to 40% by mass; and*
- *Asbestos content greater than 40% by mass.*

Results that specified more precisely must be provided with a reasoned statement on the uncertainty

F. Karkori and A. A. Olsen, *Inventory of Hazardous Materials*, Synthesis Lectures on Ocean Systems Engineering, https://doi.org/10.1007/978-3-031-76734-0

(3) *As to the asbestos types, to distinguish all six different types is time consuming and, in some cases, not feasible by current techniques; while on the practical side, the treatment of different types of asbestos is the same. Therefore, it is suggested to report the type when necessary.*

Polychlorinated Biphenyls (PCBs)

Note: There are 209 different congeners (forms) of PCB of it is impracticable to test for all. Various organisations have developed lists of PCBs to test for as indicators. In this instance two alternative approaches are recommended. Method 1 identifies the seven congeners used by the International Council for the Exploration of the Sea (ICES). Method 2 identifies 19 congeners and seven types of aroclor (PCB mixtures commonly found in solid shipboard materials containing PCBs). Laboratories should be familiar with the requirements and consequences for each of these lists.

Types to Test for:
Method 1: ICES7 congeners (28, 52, 101, 118, 138, 153, 180).

Method 2: 19 congeners and seven types of aroclor, using the US EPA 8082a test.

Specific testing technique: GC–MS (congener specific) or GC-ECD or GC-ELCD for applicable mixtures such as aroclors. *Note* standard samples must be used for each type.

Sample Preparation: It is important to properly prepare PCB samples prior to testing. For solid materials (cables, rubber, paint, etc.), it is especially critical to select the proper extraction procedure in order to release PCBs since they are chemically bound within the product. Specific reporting information: PCB congener, ppm per congener in sample, and for Method 2, ppm per aroclor in sample should also be reported.

Notes:

(1) *Certain field or indicator tests are suitable for detecting PCBs in liquids or surfaces. However, there are currently no such tests that can accurately identify PCBs in solid shipboard materials. It is also noted that many of these tests rely on the identification of free chlorine ions and are thus highly susceptible to chlorine contamination and false readings in a marine environment where all surfaces are highly contaminated with chlorine ions from the sea water and atmosphere.*

(2) *Several congeners are tested for as "indicator" congeners. They are used because their presence often indicates the likelihood of other congeners in greater quantities (many PCBs are mixes, many mixes use a limited number of PCBs in small quantities, therefore the presence of these small quantities indicates the potential for a mix containing far higher quantities of other PCBs).*

(3) *Many reports refer to "total PCB," which is often a scaled figure to represent likely total PCBs based on the sample and the common ratios of PCB mixes. Where this is done the exact scaling technique must be stated and is for information only and does not form part of the specific technique.*

Ozone Depleting Substances

Types to test for: as per appendix 8 of this text all the listed CFCs, Halons, HCFCs and other listed substance as required by Montreal Protocol.

Specific testing technique: Gas Chromatography-Mass Spectrometry (GC–MS), coupled Electron Capture Detectors (GC-ECD) and Electrolytic Conductivity Detectors (GC-ELCD).

Specific reporting information: Type and concentration of ODS.

Anti-Fouling Systems Containing Organotin Compounds as a Biocide

Types to test for: Anti-fouling compounds and systems regulated under annex I to the International Convention on the Control of Harmful Anti-fouling Systems on Ships, 2001 (AFS Convention), including tributyl tins (TBT), triphenyl tins (TPT) and tributyl tin oxide (TBTO).

Specific testing technique: As per resolution MEPC.104[49] (Guidelines for Brief Sampling of Anti-Fouling Systems on Ships), adopted 18 July 2003, using ICPOES, ICP, AAS, XRF, GC–MS as applicable.

Specific reporting information: Type and concentration of organotin compound.

Note For "field" or "indicative" testing it may be acceptable to simply identify presence of tin, due to the expected good documentation on anti-fouling systems.

Appendix J: Examples of Radioactive Sources

The following list contains examples of radioactive sources that should be included in the Inventory, regardless of the number, the amount of radioactivity or the type of radionuclide.

Examples of Consumer Products with Radioactive Materials
Ionisation chamber smoke detectors (typical radionuclides ^{241}Am; ^{226}Ra).

Instruments/signs containing gaseous tritium light sources (^{3}H).

Instruments/signs containing radioactive painting (typical radionuclide ^{226}Ra).

High intensity discharge lamps (typical radionuclides ^{85}Kr; ^{232}Th).

Radioactive lighting rods (typical radionuclides ^{241}Am; ^{226}Ra).

Examples of Industrial Gauges with Radioactive Materials
Radioactive level gauges.

Radioactive dredger gauges[19]

Radioactive conveyor gauges[56]

Radioactive spinning pipe gauges [56]

Note
[19] Typical radionuclides: ^{241}Am; ^{241}Am/Be; ^{252}Cf; ^{244}Cm; ^{60}Co; ^{137}Cs; ^{153}Gd; ^{192}Ir; ^{147}Pm; ^{238}Pu; ^{239}Pu/Be; ^{226}Ra; ^{75}S; ^{90}Sr (^{90}Y); ^{170}Tm; ^{169}Yb

F. Karkori and A. A. Olsen, *Inventory of Hazardous Materials*, Synthesis Lectures on Ocean Systems Engineering, https://doi.org/10.1007/978-3-031-76734-0

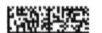